Quantification: Transcending Beyond Frege's Boundaries

Critical Studies in German Idealism

Series Editor

Paul G. Cobben

VOLUME 5

The titles published in this series are listed at brill.nl/csgi

Quantification: Transcending Beyond Frege's Boundaries

A Case Study
In Transcendental-Metaphysical Logic

By

Aleksy Molczanow

BRILL

LEIDEN · BOSTON
2012

This book is printed on acid-free paper.

Library of Congress Cataloging-in-Publication Data

Molczanow, Aleksy.
Quantification : transcending beyond Frege's boundaries : a case study in transcendental-metaphysical logic / by Aleksy Molczanow.
p. cm. — (Critical studies in German idealism, ISSN 1878-9986 ; v. 5)
Includes bibliographical references (p.) and indexes.
ISBN 978-90-04-22269-4 (hardback : alk. paper) 1. Frege, Gottlob, 1848-1925. 2. Kant, Immanuel, 1724-1804. 3. Logic, Symbolic and mathematical. I. Title.

B3245.F24M65 2012
121—dc23

2012000804

This publication has been typeset in the multilingual "Brill" typeface. With over 5,100 characters covering Latin, IPA, Greek, and Cyrillic, this typeface is especially suitable for use in the humanities. For more information, please see www.brill.nl/brill-typeface.

ISSN 1878-9986
ISBN 978 90 0422269 4 (hardback)
ISBN 978 90 0422417 9 (e-book)

If we take in our hand any volume; of divinity, or school metaphysics, for instance; let us ask, Does it contain any abstract reasoning concerning quantity or number . . .? No. Commit it then to the flames . . . .

Hume, An Enquiry Concerning Human Understanding, section12, pt.3.

That which can be destroyed by the truth should be.

P. C. Hodgel

CONTENTS

VOLUME FOREWORD

In his attempt to give an answer to the question of what constitutes real knowledge, Kant steers a middle course between empiricism and rationalism. True *knowledge* refers to a given empirical reality, but *true* knowledge has to be understood as necessary as well, and so consequently, must be a priori. Both demands can only be reconciled if synthetic a priori judgments are possible. To ground this possibility, Kant develops his transcendental logic.

In Frege's program of providing a logicistic basis for true knowledge the same problem is at issue: his logicist solution places the quantifier into the position of the basic element connected to the truth of a proposition. As the basic element of a theory of logic, it refers at the same time to something in reality.

Mołczanow argues that Frege's program fails because it does not pay sufficient attention to Kant's transcendental logic. Frege interprets synthetic a priori judgments as ultimately analytic, and thus falls back onto a Leibnizian rationalism, thereby ignoring Kant's middle course.

Under the title of the *transcendental analytic of quantification* Mołczanow discusses Frege's concept of quantification. For Frege, the proper analysis of number words and the categories of quantity raises problems which can only be solved, according to Mołczanow, with the help of Kant's transcendental logic. Mołczanow's book thus deserves its places in the series *Critical Studies in German Idealism* because it provides a further elaboration of Kant's transcendental logic by bringing it into conversation with contemporary logic. The result is a new conception of the nature of quantification which speaks to our time.

Paul Cobben, Series Editor
Tilburg University, The Netherlands

PREFACE

> In all affairs it's a healthy thing now and then to hang a question mark on the things you have long taken for granted.
>
> *Bertrand Russell*

Frege's notion of the quantifier is simply the most basic element of his programme of logicism. The downfall of the latter came only as a distant, and thus not directly traceable, result of the fundamental deficiencies of the former. However, the situation in logic appears to be such that even though the falsity of the consequence has long been recognised, the false premise is still believed to be true. In much the same way as Gödel's Incompleteness Theorem drove a wedge between the notions of what is "true" and what can be "proven," this book exposes an unbridgeable gap between the notion of "truth" expressed by the Fregean quantifier and that of "reference." In Frege's doctrine, quantificational truth as reference of the whole obtaining regardless of the reference of its components is shown as only attainable at the intolerable cost of a complete truncation of the referentiality of the latter. This monograph provides evidence that there are thus no reasons for taking the referentiality of the Frege-Russellian quantification itself for granted. The gap between truth and reference appears to be too serious to be overcome by the mere notion of "strategy" employed in game-theoretical semantics. The material necessary condition of finding a way to the solution of the problems that quantification presents to the theorist lies, thus, in the rejection of the conception of these problems as being *local* to the description of quantifiers as such, and in understanding these problems, rather, as distant and merely remote consequences of our much deeper misunderstanding of the very conception of logic; this, as it is shown in the monograph, being itself a corollary of the fact that the full theoretical relevance of Kant's basic distinction between formal and material truth, and, consequently, between general and transcendental logic, had been largely underestimated. This study provides grounds for believing that the proposed reformulation of basic logical concepts along the principles outlined in Kant's *Critique of Pure Reason* will help to arrive at an entirely new understanding of the genuine nature and the logical essence of quantification.

GENERAL OVERVIEW

> An error doesn't become a mistake until you refuse to correct it.
>
> *Orlando A. Battista*

In order to win a campaign, one has to pursue the right strategy, or one of a number of right strategies. In the latter case, any strategy will be right provided that it happens to be more effective than that adopted by the opponent. This is clearly the case in the game of chess. Then your strategy will be a winning strategy no matter what tactical moves your opponent undertakes.

This seems to be well in line with the well-known definitional strategy of Hintikka's school of game-theoretical semantics, though with certain qualifications. The point holds well only insofar these moves, however ingenious, are undertaken under a certain fixed strategy. But what happens if your opponent (Nature) turns up to be not only an excellent tactician, but a better strategist as well? She will then outplay you by choosing another strategy that may happen to be superior to yours. Then all your previous moves, no matter how brilliant they were, would fall as part of an inferior strategy.

As a corollary, a winning strategy cannot be a matter of mere technicalities, as the definitions so frequently employed by Hintikka might suggest. Merely technical solutions will definitely not do when what is at stake is a winning strategy as such, and not just winning tactics. Thus, it is a matter of an appropriate general conception, rather than one of strictly technical solutions serving no other purpose than pumping new strength into a structure that might eventually turn out to be standing on sand, no matter how promising these technical solution might seem. It does not take a specialist to see that the reliance on an illusion created by a defective idiom may quickly lead to your losing in a game. And this might be awfully frustrating, especially when the name of the game is the pursuit of truth.

The study of the phenomenon of quantification is a good case in point. Its history presents perhaps one of the most striking examples of how good tactics may conceal a profoundly misconceived strategy, with each apparent success on various tactical fronts blinding us to the inevitable overall defeat. One of the purposes of this book, thus, is not to follow the routine practice of displaying or confronting all the intricacies and articulations of the theoretical machinery within the various theoretical approaches to the study of quantification, but to investigate the general ground on which current theories of quantification may be shown to be inconsistent, all the persuasiveness of their intricacies and articulations notwithstanding. This is self-evident for, in order to know *how* to choose a better strategy, one has to know *why* the present strategy is not as good as one hoped. The best way to do it is to expose, in the spirit of Kant's Transcendental Dialectic, the roots of a philosophical illusion at the heart of current theories of quantification. The main purpose of the first part of this book, *The Transcendental Dialectic of Quantification*, is therefore to provide an in-depth examination of the sources of a widespread transcendental illusion, deeply entrenched in the minds of the students of quantification, and to show it as the result of the underestimation and misinterpretation of Kant's distinction between general and transcendental logic, owing to which the proper and unique task of the latter, that of elucidating the conditions for the possibility of any discursive thought, has almost entirely disappeared from view. As Alexander Levine rightly observes, in the time that followed the publication of the *Critique of Pure Reason*, the distinction in question was, quite expectedly, blurred or deliberately abandoned by nineteenth-century German idealists and British empiricists alike,[1] with this naturally expected result also leading to a subsequent rejection of, or else complete ignorance of the need for, the required winning strategy that Kant had laid out in his unequalled philosophical classic.

As a strategic thinker, Kant has been rightly recognised as the architect who provided conceptual design sketches for the new edifice to be built on the site once occupied by syllogistic logic.[2] Kant's *Critique of Pure Reason*, like an architectural blueprint, offered, indeed, the most general plan for the new construction yet to be accomplished. But it would be too hasty to assert, along with some philosophers, that Frege, Hegel, Russell, Gödel and others would—according to Kant's plan—do the actual technical engineering work necessary to erect the new building on the site once occupied by Aristotelean logic but which by the eighteenth century was reduced to the rubble left by Ramist and Cartesian demolition gangs. Early analytic philosophy, to the contrary, developed largely in reaction to Kant, *not* as a continuation of work along Kant's guidelines. Unlike Frege

[1] A. Levine, *Scientific Progress and the Fregean Legacy*, p. 266.
[2] M. Tiles, *Kant: From General to Transcendental Logic*, p. 85.

and Russell, who besides being known as precursors of analytic philosophy were also profound followers of Leibniz and, therefore, manifestly rationalist thinkers, Kant's strategy was to chart a sure course between the Scilla of Leibniz's Rationalism and the Charybdis of Hume's Empiricism. He was aware that taking the right course between them was a question of life and death for metaphysics, as adopting the principles of the latter was irreconcilable with the very existence of metaphysics as such, while those of the former were bound to make metaphysics of no more significance than any barren exercise. All this notwithstanding (and even despite his expressed agreement, in *Die Grundlagen der Arithmetik*, with Kant on the synthetic *a priori* status of truth in geometry), Frege split the horns of Kant's dilemma by contending that all synthetic *a priori* is merely analytic (the reason behind taking this move was apparently the need to restate Leibniz in his own terms, keeping the rationalist content intact so as to secure his *Begriffsschrift* as a rightful version of Leibniz's characteristic language). That was the starting point of Frege's *logicist thesis*, as well as the first step toward the failure of Frege's entire programme (which likewise turned out to be a source of grave problems in early analytic philosophy that subsequently led to the downfall of the ideal language conception of logical atomism).

Acting out of loyalty to his truly Leibnizean project, Frege dismissed the considerations that had led Kant to the full appreciation of the philosophical significance of the problem of *a priori synthesis* as being the most pressing question in the entire field of speculative philosophy. For Frege, the analytic was still interchangeable with *a priori*, and the synthetic with *a posteriori*, quite regardless of Kant's prior discovery that the *a priori* and the synthetic do not exclude one another. Thus, Frege's main thrust behind the advancement of his foundational Law or Axiom V (which, as an axiom, was supposed to need no proof, yet the truth of it as an axiom was not entirely obvious even to Frege himself) was to provide substantiation for his thesis that any judgment even if it appeared as synthetic was, at bottom, analytic.[3] Even though toward the end of his life Frege himself came to realise that this enterprise had been profoundly misconceived, he no longer had time or energy to look for, or enquire into any alternative solutions. For Kant, on the contrary, the problem that, even if not

[3] "The conclusions we draw from it extend our knowledge, and ought therefore, on Kant's view, to be regarded as synthetic; and yet they can be proved by purely logical means, and are thus analytic." G. Frege, *The Foundations of Arithmetic*, p. 101.

yet fully resolved *in concreto*, was at least rightfully addressed, was "How are synthetic *a priori* judgments possible?" Roughly, the question boiled down to the enquiry into what are those elements in knowledge that render possible the relations constitutive of *meaning*. Kant, however, never attempted to elaborate this into a complete *logical* doctrine. Being well aware that he had touched upon the most important issues in the *entire* field of philosophical enquiry, he hasted to apply his new Critical principles in the spheres of morality, aesthetics, and teleology, thus evidently regarding it quite dispensable, at the time of his writing, to make yet further effort in securing his propaedeutic to a system of pure reason—that forms, as it were, a vestibule of transcendental logic—by a full in-depth enquiry into that discipline.

On the face of it, the absence of any subsequent full in-depth enquiry into matters that Kant regarded as having been established as a matter of principle, if not *in concreto*, fueled various skeptical responses. These ranged from those that asked whether, at worst, it is even possible to exhibit his theory as a coherent, unified scheme to those that concluded that, at best, it draws an imperfect picture, the accuracy of which has not been fully ascertained. What, in the wake of Kant's Copernican revolution, seems however to be beyond any doubt is that in much the same way as true astronomy requires, to explain the movement of the heavens, the recognition of the movement of the observer—true logic, in order to achieve its aim, also requires the recognition of the substantial role played by the mind that is able to entertain thought without accepting it. Viewed in the light of this paradigm-changing perspective, thought is no longer merely "grasped" by mind in some mysterious manner that, according to Frege, is of no interest for, and thus no business of, logic. Rather, mind appears to be playing a thoroughly constituitive role, in which it is mind that ultimately crafts its own patterns to suit mind's own needs and capacities, instead of getting the ready-made stuff in the form of mysterious Platonic entities straight from the wholesalers of the Fregean third realm. Kant was correct when he saw that human experience was not atomistic, as Hume had thought, but instead was permeated by *a priori* structures; yet Kant's formulation of those structures, reflecting his complete belief in Newtonian physics, was inevitably too narrow and simplistic.

In view of this, the pressing need for further enquiry into the subject matter of Kant's system of pure reason becomes evident. This task is undertaken in the second part of the monograph, with the range of issues discussed in *The Transcendental Analytic of Quantification* being necessarily confined to the topics centering around the main issue of Frege's polemic with Kant;

namely whether truth in arithmetic is analytic, unlike geometry where it is synthetic. In this polemic, of the utmost significance to the choice of the topic of our enquiry is the evolution of Frege's views on number and language, in the process of which he finally drops his claim, against Kant, that "nought and one are objects which cannot be given to us in sensation"[4] and comes to an entirely opposite conclusion that "the numerals are absolutely *not* proper names or objects." To this he adds:

> So that a sentence like 'four is a square number' does not express the subsumption of an object under a concept. And so this sentence cannot at all be construed like 'Sirius is a fixed star'. But then, *how*?[5]

It will be evident that even though from this conclusion of Frege's it does not necessarily follow that he already wholly sides with Kant, there is yet one moment, or undertone, that places Frege and Kant in one line. This happens to be Frege's explicit concession, in the passage above, that the proper analysis of number words and the categories of quantity with which they are but most intimately connected, yet remains an open problem, a question that needs further and deeper investigation. Since, however, all previous formal logic, and thus general logic in Kant's sense, has shown a critical lack of the appropriate theoretical resources even in principle, the only proper framework for this problem to be dealt with is but that of transcendental logic, construed and elaborated further strictly along Kant's guidelines. Therefore, the second part of the book pursues a task of inquiring specifically into the transcendental logic of quantification in an attempt to discern and elucidate more fully, upon the lines and in the spirit of Kant's transcendental philosophy, the place and the role of the categories of quantity in Kant's synthetic *a priori*.

The present investigation need not be anachronistic, however, in the sense that acting along the lines and in the spirit of Kant's transcendental philosophy does not imply that this study should be kept entirely in the orbit of philosophical thought that existed at the time of Kant's writing. Since the study focuses on the central question of transcendental logic, namely the question regarding the ground of reference to objectivity posed by Kant in his famous 1772 letter to Herz, but to which he however had found neither time nor the means to provide a fully comprehensive

[4] G. Frege, *The Foundations of Arithmetic*, p. 101.
[5] G. Frege, *Nachgelassene Schriften*, p. 282.

in-depth solution,[6] it necessarily draws on the solutions and theoretical resources that would be found in logico-philosophical investigations after Kant, like the modern notion of Skolem function, Ch. S. Peirce's Type/Token/Tone distinction as well as his known thesis regarding the impossibility of thinking without signs, or Wittgenstein's negative solution to the problem of 'private language'. All this is not aimed at whatever revision of Kant's philosophical heritage, however; the task of the present study is simply to make an attempt to arrive at more substantial results in the grossly unexplored field of transcendental logic as a science *in concreto* in a sense none other than Kant's own.

[6] "I noticed that I still lacked something essential, something that in my long metaphysical studies I, as well as others, had failed to pay attention to and that, in fact constitutes the key to the whole secret of hitherto still obscure metaphysics. I asked myself: What is the ground of the reference of that in us which we call "representation" to the object?" Immanuel Kant, *Philosophical Correspondence*, pp. 129–30.

THE TRANSCENDENTAL DIALECTIC OF QUANTIFICATION

CHAPTER ONE

THE FAVOURED DISTINCTION

> The more genuinely a methodological concept is worked out and the more comprehensively it determines the principles on which a science is to be conducted, all the more primordially is it rooted in the way we come to terms with the things themselves, and the farther is it removed from what we call "technical devices", though there are many such devices even in the theoretical disciplines.
>
> *M. Heidegger, Being and Time, p. 50.*

> This is an exemplary case of running hard to stay in the same place. Here again the necessity for the favoured distinction [...] is a consequence of an antecedent theory, not of the natural phenomena of language. If the planets revolve around the earth on crystalline spheres, we must investigate of what adamantine crystal the spheres are made. But we might instead question the premise.... It is *this* that should be questioned and investigated.
>
> *G. P. Baker, P. M. S. Hacker,*
> *Language, Sense & Nonsense, p. 115.*

1.1. *Foundational Goals—Strategy and Tactics*

Everyone versed in modern logic knows that it originated with Frege, who "clearly depicted polyadic predication, negation, the conditional, and the quantifier as the bases of logic".[1] The quantifier was a central item in the logical systems of both Frege and Russell, and it is still an all-important notion of modern symbolic logic. It is quite unsurprising that the modern version of Fregean logic has not only become the backbone of all currently developed systems of logic but has exerted its overwhelming influence on modern linguistics as well. In view of all this, the concept of the quantifier seems to have become part of the basic strategy of both logicians as well as linguists.

[1] W. Goldfarb, *Logic in the Twenties: the Nature of the Quantifier*, p. 351.

Most recent developments in logic, namely the advancement by Hintikka of the conception of independence-friendly logic (IF-logic), claimed by him to be both a conservative extension as well as the foundational region of ordinary (Fregean) logic (grounded in the distinction between dependent and independent quantifiers, which he sees as basic for our understanding of the real situation in logic),[2] leave no doubt that the received notion of the quantifier as the Fregean expression of generality has long been upgraded to the rank of a central item as well as most important strategic tool in modern inquiries into the nature of language and thought.

Truly, Frege was the founder of modern symbolic logic. At the time of his writing, it was an entirely new conception of logic, the novelty of which was so great that it made it hard for his contemporaries to recognise the full significance and the whole rich array of consequences following from his work. Yet, the real assessment of the novelty of what has now been hailed in many ways as Frege's revolutionary achievements lies not in the purely technical results that he obtained, but in something else that, in a sense, existed prior to these results and, just owing to its preexistence, made these results possible.

As for the results themselves, it would be hard to tell, as is also the case with Hintikka's IF-logic, to which extent it belonged to the truly foundational region, and to which it was just a conservative extension of the results of his predecessors. For among the four items listed by Goldfarb as the bases of Frege's logic, negation and the conditional had already been part of a long-standing tradition in logic. Having its roots deeply grounded in propositional logic, Frege's logic could be rightly seen as a truly conservative extension of the classical calculus of propositions. Frege's polyadic predication, too, can be regarded as a conservative extension of the archetypal theory of monadic predication, the latter turning up to be a particular case of the former in exactly the same way as Frege's own logic is being shown, by Hintikka, to be no more than just a particular case of his IF-logic. It would be more promising, however, not to trace the ways in which one's theory might be regarded as a conservative extension of someone else's theory, but to see why it is so that, of the two, a theory and this theory's 'conservative' extension, it is exactly the latter which

[2] J. Hintikka, *The Principles of Mathematics Revisited*, Cambridge, 1996.

happens to be forming the foundational area for the former, and not vice versa.

It is fairly evident that what makes one theory a conservative extension of another is its incremental character with respect to the latter, in the same way that it is clear that what makes a theory more incremental is just the extent to which it is more abounding in technical detail with respect to another. What makes one theory to be an extension of another may be seen, then, as a purely mechanical question of expanding the theory's nomenclature. This, in effect, is a merely tactical question: what are the items off the shelf of the technicalities available that we should choose in order to make the expansion of the theory's nomenclature worthwhile. But this alone would not suffice to see how and why these incremental parts should acquire quality that is truly foundational. What is much more interesting and more essential, thus, belongs not to the tactics, but to the strategy of theoretical inquiry; namely, not exactly *what the items* in question are, but on *what principles* we choose these items but not others. In short, it is not *what* we choose but *how* we choose (no matter what) that is crucial and much more important for seeing the real course of things.

To analyse the more primordial region of theory construction we must, hence, drop the idea of treating purely technical devices such as the quantifier as basic strategic means of attaining the goals of theoretical research. In fact, they are no more than merely tactical technicalities serving for the attainment of certain tactical goals and, as purely tactical means, they are no part of strategy. A strategy is something that exists prior to any of the tactical moves, and these may indeed be very different even when the strategy remains the same.

What both Frege and Hintikka appear to have in common in this primordial sense is a uniform strategy, and this strategy would be best understood if we first put under scrutiny the strategic goals, in fact, the same strategic goal that they happen to share. This is exactly the goal which Hintikka professedly attains from the principles on which his IF-logic happens to be based—one of attaining *descriptive completeness*. And this is also the goal that Hintikka's famous predecessor had tried to achieve by introducing his gothic letters which he placed in a concavity on his content-stroke. No wonder, then, that the strategies that they choose for attaining this goal, even though differing in minor details, both deal with the same technical device and both happen to be based on the same strategic lines.

The same technical device is, of course, Frege's technical expression of generality, now known as the Fregean quantifier.

1.2. *Natural Language vs. "Formalised Language of Pure Thought"*

The goal that Frege strived to achieve was to invent a symbolic substitute for natural language, the substitute which he called "a formalised language of pure thought," whose task was to serve as a perfect tool for checking the cogency of logical proofs. It is interesting to note that a language is essentially a set of all sentences in *L*, where *L* is the same very language, and, in one of the senses in which the term 'theory' is being understood, so is a theory itself. Since, in this formal sense, 'theory' and 'language' are interchangeable, we can therefore speak of either descriptive or else deductive completeness of a theory or, alternately, of a (formal) language of this theory, having in both cases essentially the same thing in view.

One interesting moment to be noted here is that designing a formal theory may proceed regardless of the specific content of this theory. Different senses of completeness, noted by Hintikka in his *The Principles of Mathematics Revisited,* namely completeness in the descriptive sense and completeness in the deductive sense, thus apply to the content of Frege's *Begriffsschrift* only indirectly, via its formal structure due to its one-to-one correspondence to the logical structure of 'pure thought' of which *Begriffsschrift* was thought of as an intended model (*modulo isomorphism*). Therefore, in order to be able to perform the intended function of an appropriate tool for checking the validity of logical proof, i.e., to be *deductively complete,* Frege's 'formalized language of pure thought' had, in the first place, to be *descriptively* complete and, thus, it had to represent directly—and in this way presuppose some sort of previous existence of—a formal theory of logical structure. In other words, Frege's task of devising *Begriffsschrift* as a language which would, unlike natural language, be able to mirror logical structure in a one-to-one fashion necessarily involved or, rather, inevitably presupposed a more foundational aspect of his purely technical enterprise, namely the task of devising a theory of logical structure as such. Frege himself emphasised this foundational aspect of the construction of an ideal logical language in one of his posthumously published writings:

> If our language were logically more perfect, we would perhaps have no further need of logic, or we might read it off from the language. But we are far from being in such a position. Work in logic is, to a large extent, a struggle

> with the logical defects of language, and yet language remains for us an indispensable tool. *Only after logical work has been completed shall we possess a more perfect instrument.*[3]

In this, he followed the lead of Trendelenburg who, in his paper 'On Leibniz's Project of a Universal Characteristic' (which Frege apparently read), had recounted Descartes' critical remark on such undertakings to the effect that "the invention of such a language depends on the true philosophy" and noted that Leibniz's project had remained incomplete since the necessary philosophical analysis of concepts had not yet been achieved.[4]

Since Frege worked on the assumption that "languages, whether natural or invented, may accordingly differ in respect of their capacity accurately to mirror thought-structures in sentence-forms,"[5] and in *Begriffsschrift* he conceived of concept-script "as one language among others (special only in meeting the standards of logical perspicuity),"[6] it was only natural that he regarded all related theories of sentence-forms as generating inaccurate, distorted representations of the conceptual structure of 'pure thought'. The most adequate or else least deviant representations he found in the formula languages of mathematics which, as he noted, "come much closer to this goal, indeed in part they arrive at it,"[7] and, specifically, in the formula language of arithmetic—due to its ability to form "concepts of such richness and fineness in their internal structure that in perhaps no other science are they to be found combined with the same logical perfection."[8]

In Frege's view, however, the language of arithmetic was only lacking one of the two main components which any language must possess—"the formal part which in verbal language comprises endings, prefixes, suffixes and auxiliary words", so its logical perfection was inevitably confined solely to the "material part of a language."[9] Thus, the language of arithmetic was, in a sense, incomplete, since the logical perfection in

---

[3] G. Frege, 'My basic logical Insights,' in *Posthumous Writings*, (Hermes, H., F. Kambartel, F. Kaulbach, eds., Basil Blackwell, Oxford, 1979), p. 252.

[4] A. Trendelenburg, 'On Leibniz's Project of a Universal Characteristic', *Historische Beiträge zur Philosophie*, vol. 3, Berlin, 1867. Cited from: H. Sluga, *Gottlob Frege*. Routledge & Kegan Paul, London, 1980, p. 51.

[5] G. P. Baker, p. M. S. Hacker, *Frege: Logical Excavations* (Basil Blackwell, Oxford, 1984), p. 66.

[6] *Ibid.*, p. 212.

[7] G. Frege, 'Boole's logical Calculus and the Concept-script,' in *Posthumous Writings*, (Hermes, H., F. Kambartel, F. Kaulbach, eds., Basil Blackwell, Oxford, 1979), p. 13.

[8] *Ibid.*, p. 13.

[9] *Ibid.*

question no matter how highly appraised applied only to the content, but not to the form and, thus, only to the 'lexicon', but not logical 'grammar' of the intended language of pure thought. In this way, the language of arithmetic appeared as a language of a logically perfect lexicon with a logically imperfect grammar. So, for Frege, the problem arose of devising signs for logical relations—a sort of grammatical 'form words'—"that are suitable for incorporation into the formula-language of mathematics, and in this way of forming—at least for a certain domain—a complete concept-script."[10] This was a necessary as well as sufficient condition for the mathematical theory of magnitudes to meet the broader requirements of the theory of the structure of 'pure thought', since arithmetic, as Frege noticed, "only forms numbers out of numbers and can *only* express those judgements which treat of the equality of numbers which have been generated in different ways."[11] In a way, to Frege arithmetic appeared as the correct—though restricted only to a part of the intended model—theory of logical form such that, in Hintikka's terms, it may be said to be descriptively incomplete. And the reason of this descriptive incompleteness of the language of arithmetic lay exactly in the complete absence of what Hintikka would have seen as deductive completeness.

In order to achieve the required deductive completeness of his concept-script, Frege provided the supplementation of "the mathematical formula language with signs for logical relations, so that initially a *Begriffsschrift* for the area of mathematics should come out of it."[12] However, Frege was fully aware of the fact that a mere combination of equations and inequalities of arithmetic with his graphical signs for his basic logical relations (negation and the conditional) was simply a technical (if not purely mechanical) enterprise which did not yet yield the required quality of a *philosophical* language to his concept-script and, thus, did not qualify it as *descriptively* complete. He saw clearly that the attainment of that goal depended on whether

> The use of my symbols in other areas is not excluded through this. . . . Whether this happens or not, the intuitive representation of the forms of thought has in any case a significance that goes beyond the area of mathematics. For that reason, may philosophers pay some attention to this matter.[13]

---

[10] *Ibid.*, p. 14.
[11] *Ibid.*, p. 13.
[12] G. Frege, 'Über die wissenschaftliche Berechtigung einer Begriffsschrift', p. 113.
[13] *Ibid.*, pp. 113–14.

To see whether the attainment, in Frege's *Begriffsschrift*, of what Hintikka would later call *descriptive completeness* was real or only imaginary, it seems necessary to give a bit more scrutiny to the specific foundations on which Frege grounded both his strategy and his belief that his task of constructing a truly *philosophical* language had, indeed, been completed.

1.3. *Grammar vs. Language: The Quest for Basic Distinction*

As Frege was fully aware, the construction of a truly philosophical language required previous knowledge of a truly philosophical grammar.[14] For Frege, the traditionally known distinctions (and thus categories) on which the grammars of Indo-European languages were based (among these, also the distinction between subject and predicate) were logically defective and thus they could not form the basis of a grammar of 'pure thought'. To arrive at this basis (and thus to attain an adequate *philosophical* conception of logic *per se*), Frege had to look for an appropriate distinction that would be fruitful as well as necessary for the description of truly *philosophical* grammar. In view of the putative 'logical perfection' of the formula language of arithmetic, for the reason of which it was selected, by Frege, for the model of his 'language of pure thought', it was only natural to think that the distinction in question could only be grasped in the specific character of "the symbols customarily used in the general theory of magnitudes [which] fall into two kinds,"[15]—the point that Frege made, notably, at the very start of the first paragraph of his *Begriffsschrift*.

The distinction between constant symbols and variables (or, in Frege's terminology, letters) that Frege took over from the formalised language of arithmetic was exactly the distinction on which he rested the whole edifice of his philosophical grammar. The paramount importance of this distinction followed, for Frege, from the fact that in algebraic expressions like $(a + b) \cdot c = a \cdot c + b \cdot c$ which Frege consigned to sentences expressing generality, "we have the ordinary use of letters to indicate."[16] As Frege explains,

> They serve here to confer generality on the thought. They stand in the place of proper names, but are not such (pronouns). You always obtain a

[14] For the explanation of the terms "philosophical" or "universal" grammar, very poorly known today, see N. Chomsky, 'Language and Mind', p. 13 ff.

[15] G. Frege, *Conceptual Notation and Related Articles* (Clarendon Press, Oxford, 1972), p. 111.

[16] G. Frege, 'Logical Defects in Mathematics', *Posthumous Writings*, p. 162.

> particular case of the general sentence when you substitute the same proper name of a number throughout for the letter *a*, and similarly for *b* and *c*. The sentence now says—and just this is what constitutes its generality—that a true thought is expressed in this way no matter what proper names of numbers may be substituted.[17]

Since, however, it is the use of constant symbols alone, namely the juxtaposition of constant symbols (names of operations) with other constant symbols (proper names of numbers) that would deprive the algebraic expression '$(a + b) \cdot c = a \cdot c + b \cdot c$' of its status of an expression of generality, and it is the study of generality that happens to be of special interest for logic, it becomes clear that the key to the most general way of expressing generality and, thus, the key to the construction of the grammar of a philosophical language lies exactly in the juxtaposition of these two *different* kinds of symbols—constants and variables. What becomes a matter of prime importance for Frege is thus not the issue of the mere existence of constants or variables as such, but the issue of the pre-existence of the *difference* between constant and variable signs that makes the distinction between constants and variables possible, and which the language of mathematics only makes explicit. The required universality—and in this way also the status of a *philosophical* language—of his concept-script now seems to be attainable, provided that one construes its grammar on the basis of this categorial distinction.

It is of crucial importance now to bring to light *how* and in exactly *what way* drawing the distinction in question can be seen to provide the principled basis for viewing Frege's system of logical notation as a language of pure thought 'from which logic might be read off directly'. It is pretty evident that whatever is of importance here lies not so much in the mundane observation of the fact that variable signs in algebraic formulas serve to *confer generality* on the thought that these formulas express. For Frege this serves the purpose of a rhetorical device of no more than presentational significance. What is much more important—indeed essential—but of which he never speaks quite explicitly is not just a mere statement that *variables confer generality*, but its premise, from which this statement follows only as its consequence. For what may also be observed is that it is only in certain circumstances, namely in the *presence* of variable signs that *constant signs*, too, may be seen as acquiring *generality*. And this opens the way for upgrading *constant signs*, not variables, to the rank of

---

[17] *Ibid.*

logical constants—in other words, they immediately qualify as elements of the logical grammar of Frege's concept-script and can thus be relegated to its 'formal part which in verbal language comprises endings, prefixes, suffixes and auxiliary words'. The next step in the construction of 'the formal part' of his concept-script will, then, amount to improving on the functional generalisation of arithmetical or algebraic expressions so that this functional generalisation can now become the truly logical grammar of his logical language. Frege's strategic goal was then to find an appropriate *extension* of function theory, namely to demonstrate that functional representation could be now applied as a mode of presentation of logical entities and, thus, of such entities which had not been shown, at least not before Frege, to fall into the descriptive and hence explanatory purview of mathematical function theory.

1.4. *Extending Function Theory*

What occupies much of Frege's published as well as unpublished work concerned with the elucidation of the workings of his concept-script and, correspondingly, gains respective prominence in the publications of the various commentators on Frege is the discussion of different variations of the relation of *falling* (of an object *falling under* a concept or of a concept *falling within* another concept of a higher level) and of how this relation is perfectly matched by his conceptual notation. This, indeed, may be seen as significant since, starting with showing how expressions like $2^4 = 16$ can be seen as depicting mathematical objects like *the number* 2 as falling under the mathematical concepts like *the 4th root of 16*, we can now show how mundane objects can be seen as falling under mundane concepts. In other words, we can then extend the functional representation of '$x^4 = 16$' in terms of its functional generalisation in $F(x)$ by showing that this generalised notational variant, $F(x)$, may also be applied to cases in which objects *other than numbers* can be seen as falling under respective concepts. But in Frege's writings this, too, appears as a rhetorical device of presentational significance only: the aim that he strives to achieve through this kind of demonstration is to show *how* his concept-script could be accommodated for the purposes of other disciplines, but—paradoxically enough—this does not mean that he also arrives at the demonstration of *whether* this can be done at all. For Frege never gives indications that he considers his exegetic manoeuvres as demonstrations of something that has the status of an axiom and thus requires no proof whatever. On the contrary, as can be seen in the passage already cited, Frege states quite explicitly that *whether this happens or not* (i.e. whether objects *other than numbers* can also be seen as falling under respective concepts), his representation of the forms of thought has *in any case* a significance that goes beyond the area of mathematics.[18]

The key to this apparent paradox can be seen in the remark that he adduces: "For that reason, may philosophers pay some attention to this matter."[19] Since Frege is specifically concerned with devising a concept-script for the purposes of *mathematics*, that is supposed to be the main area of its application. The falling of objects other than numbers under the respective concepts appears to him as a matter of only marginal interest. And, as we see from his remark, for the purpose of constructing the philosophical grammar of his concept-script, this, of course, is desirable but not essential. What is of crucial importance, however, is to find the ways in which the constant/variable distinction that he borrows from function theory serves as a means of plain representation of logical relations *within* the area of mathematics.

Frege is generally assumed to have resolved this task via the introduction of a special sort of function, namely one taking from objects to truth values. But, even though this is considered to be one of the most important ingredients of his new logical doctrine, this kind of innovation, taken alone, was only of marginal importance, for the attainment of a truth value for Frege's functional expression necessarily presupposes that the argument places in this expression should be already filled by the proper names of respective objects.[20] And their proper names appear to be constant, not variable, signs. This operation works, therefore, only outside the distinction that Frege puts in the basis of his logic and again may be seen as a marginal device, thus, a device of presentational significance only. The basic task in arriving at a universal philosophical language applicable for the area of mathematics thus cannot be resolved solely by allowing, in his notation, objects other than numbers to fall under a concept, provided that these objects, like numbers, are considered as individuals. In fact, extending the domain of objects that are admissible as values of argument-places in a functional expression leaves the status of *Begriffsschrift* intact: in this respect it remains identical to the language of arithmetic, for

---

[18] G. Frege, 'Über die wissenschaftliche Berechtigung einer Begriffsschrift', p. 114.

[19] *Ibid.*

[20] ... what Frege's expression, $\vdash 2^4 = 6$, shows quite explicitly in the context of his example cited on p. 16 below.

the simple reason that the status of a universal language, by itself, gives no warrant that the language in question is also philosophical. In the combination of these two attributes of language *universal* comes as a corollary of *philosophical*, whereas reasoning the other way round can apparently be seen as a manifest fallacy.

Thus, much of Frege's explicit talk of logically foundational aspects of his *Begriffsschrift* appears to hit purely tactical grounds, but does not—and could not—inform his basic strategic guidelines. The attention of the reader was always drawn—for purely demonstrative purposes which were none other but illustrative—to the same, non-relational, kind of entities. However important for the readers of Frege's works and for the modern commentators on Frege,[21] his expositions revolving mainly around the issue of objects other than numbers falling under a concept in the functional representation of the latter did not, nevertheless, touch upon the issue which could be rightly said to belong to the most important aspects, indeed, to the heart of his logical doctrine.

For entirely expository purposes, and thus led by purely tactical considerations, Frege preferred to speak about objects other than mathematical ones filling the argument places of his functional representations. His deliberations, therefore, revolved around what might have seemed as beneficial for Frege's tactical purposes but were, at root, grossly inessential aspects of the issue of the extension of function theory: they hardly touched upon the aspect of this extension which was of utmost importance for Frege's underlying philosophy. Rather, they centred on what Frege might have thought to be most expedient for his potential readers to know about his doctrine in order to gain at least elementary understanding of his logical insights. Couched in purely expository terms, they reveal his tactics but they turn up to conceal rather than reveal his basic strategic line. What Frege was really concerned with in his strategic plan of devising a Begriffsschrift modelled on the formula language of arithmetic was, in effect, something different. Namely, the construction of his Begriffsschrift required such an extension of the notion of falling under a concept so that entities of a kind entirely different from mere individuals would be allowed for the values of argument-places for functions, quite regardless of whether these individuals were regarded as mathematical or non-mathematical ones. The elaboration of his philosophical grammar had an absolutely necessary pre-requisite, and this was the necessity

[21] Cf. Dummett (1973, 1981), Bell (1979), Sluga (1980), Currie (1982), Burge (1986).

of finding the mode of presentation of quite specific entities—outside mathematical function theory—that were not individuals (objects or truth-values) but relations. And the crucial point of this programme was concerned with the issue which, for him, was of utmost importance: how and exactly in what way would it be possible to incorporate the constant/variable distinction in question into the representation of relations involving logical generality?

1.5. *The True Basis of Frege's Logic: Function or Relation?*

The way Frege resolved or rather held that he had resolved this problem deserves, thus, a bit more serious scrutiny. This can be done more perspicuously if we resort to Frege's own distinction between what he calls the material and the formal part of a language, in fact any language, be it ordinary (vocal) language or its artificial logical transcription, existing primarily in a notational form (and thus having no necessary connection with whatever vocal manifestations it may find). To put it in linguistic jargon, the distinction that Frege has in mind is essentially the one between the grammar of a language and its lexicon. This, however, is the only distinctive parameter of his language of pure thought that he draws from the received linguistic tradition. For Frege's purposes of perspicuity the excessive verbiage of spoken language is something that he wants to escape, and in view of his mathematical objectives the task is easily attainable: this becomes smoothly replaced by the already existing mathematical notation. For Frege's parallel task—one of attaining generality—he applies the analysis in the terms of the function-argument distinction rendered by the theoretical apparatus of function theory. But the generality that Frege thus obtains is valid only insofar as *lexical* generality might be concerned, since the structural description that Frege imposes on, or applies to, the 'already existing' expressions of his *Begriffsschrift* is so far applicable only to its morphology. But, in this way, its grammar is incomplete. What Frege definitely needs in order to accomplish the construction of his *Begriffsschrift* is to apply the basic distinction of an antecedent theory directly to the syntactic component of his logical grammar.

The novelty that Frege introduces here is his rejection, as the basis of his grammar, of the juxtaposition of concepts—the main preoccupation of traditional logic which he sees as being based on a presumably misleading analysis of spoken language—the main reason for which he also rejects the traditional distinction between subject and predicate which he is assumed, mistakenly, to have replaced by the distinction between the variables and constants of function theory. What he puts in the basis of his grammar is not the relationship between two general concepts described in terms of the purely grammatical relationship between the subject and the predicate, which he discards as logically irrelevant, but a purely logical relation which appears to be of utmost importance for the study of logic—that of logical entailment. This, in fact, happens to be part of his general strategy of eliminating *primary propositions* (designating the relation of concepts) by translating them into *secondary propositions* (designating the relation of judgements).[22] It is improper to say, then, as we are often told, that Frege's main novelty in logic amounted to the substitution of the main logical distinction traditionally coined in terms of subject and predicate by the purely mathematical distinction between a function and its argument. For the latter is, for Frege, a fact which has nothing to do with the expression of the relationships between concepts, i.e. with the subject-predicate relationship as such, which refutes traditional

[22] ...and thus completely eliminating *primary logic* as a study of the relationships between concepts by converting it into a fragment of *general logic*, the latter being exclusively concerned with relations between propositions as such. It is exactly this that leads Heidegger, in his report on "Recent Research in Logic" (1912) as well as his dissertation (1914), to question the philosophical importance of (what he refers to as) "G. Frege's logico-mathematical investigations". As he puts it in his dissertation, "It would have to be shown how its formal nature [that of symbolic logic] prevents it from gaining access to the living problems of the meaning of propositions, of its structure and cognitive significance," since, as Heidegger stresses it in his report,

> ... it seems to me that above all it must be pointed out that symbolic logic never gets beyond mathematics and to the core of logical problems. I see the barrier in the application of mathematical symbols and concepts (above all the concept of *function*) whereby the meaning and shifts in meaning in propositions are obscured. The real significance of the principles remains in the dark; the propositional calculus, e.g., is a figuring with propositions; the problems of the theory of propositions are unknown to symbolic logic. Mathematics and mathematical treatment of logical problems reach limits where their concepts and methods fail; that is precisely where the conditions of their possibility are located.

Equally characteristic is also the following passage from *Nietzsche*, 2 vol. (Pfullingen, 1961), 2: 487:

"The signal of the degrading of thinking is the upgrading of symbolic logic to the rank of true logic. Symbolic logic is the calculative organization of the unconditional ignorance regarding the essence of thinking provided that thinking, being thought essentially, is that creative knowledge of fundamental outlines which, in the care of the essence of truth, rises from being." (Translated by Joan Stambaugh in *The End of Philosophy*: New York, 1973, p. 80; cited from A. Borgmann, "Heidegger and Symbolic Logic," in *Heidegger and Modern Philosophy*, M. Murray, ed., Yale University Press, New Haven and London, 1978, pp. 6–8.)

demonstrations of the alleged logical relevance of Frege's constant/variable distinction, cf.:

> We must likewise distinguish the case of one concept being subordinate to another from that of a thing falling under a concept although the same form of words is used for both. The [following] examples . . .
>
> (5.1) $\begin{array}{l} \vdash\!\!\!\top\!\!\!-\; x^4 = 16 \\ \quad\;\llcorner\; x^2 = 4 \qquad \text{and} \qquad \vdash 2^4 = 16 \end{array}$
>
> show the distinction in the concept-script.[23]

It is exactly 'the case of one concept being subordinate to another' that Frege speaks about as being depicted by the traditional logical division into subject and predicate, and clearly not his case of 'a thing falling under a concept'. There is absolutely no ground, then, to suggest (even though this is a generally accepted point of view) that the crucial element of the innovation that Frege produced in logic amounted to the replacement of the old distinction between subject and predicate by the modern distinction between the predicate and a varying number of its arguments, which is just one in the case of monadic predication (what—as we are often told—is exactly as in Aristotelian logic) or more than one as in Frege's so-called polyadic predication (allegedly a great advance in expressive means over the exclusively monadic predication of old logic). For, in the passage above, Frege himself in fact provides nothing else but a clear demonstration that, "the same form of words" notwithstanding, the monadic predication of Aristotelian logic is logically different, in a very essential way, from what is being understood in the same way (for the same term is used) as the monadic predication in logic after Frege. Rather, the real essence of Frege's innovation is *not* that he *extends*, as the basis of his logic, the notion of predication so as to include cases of polyadicity which old logic was unable to construe but that he completely *eliminates* the monadic basis of old logic and introduces, *instead*, the new basis which is *entirely* polyadic. Thus, if the basis of his logic were to be viewed as an extension of function theory, from which he started, then this extension would have to be seen as being formed not by a function but by a *relation*.

A function, no doubt, may be viewed as a special kind of relation, but the relation that Frege takes as the basis of his logic will be seen as making absolutely no sense as a function. To obtain the widest scope of indeter-

---

[23] G. Frege, *Boole's logical Calculus and the Concept-script*, p. 18.

minateness in meaning, and hence, the required generality of its logical meaning, Frege forms his sign—which he takes as basic because he uses it as a sign of the most primitive, and hence basic logical relation—as the designation of the negation of 'not *A* and *B*', so that his sign

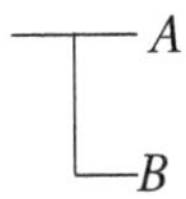

comes to be standing as a common designation for set *R* of the other three possibilities, '*A* and *B*', '*A* and not *B*', 'not *A* and not *B*', namely:

$R = \{(B, A), (\bar{B}, A), (\bar{B}, \bar{A})\}.$

In set-theoretic terms, both functions and relations are sets of ordered pairs, but relations $(\bar{B}, A)$ and $(\bar{B}, \bar{A})$ in set *R* are not functions. In a function, no two ordered pairs have the same first component, which is '$\bar{B}$' for both $(\bar{B}, A)$ and $(\bar{B}, \bar{A})$. In order to be an extension of functional representation, set *R* should therefore contain either $(\bar{B}, A)$ or $(\bar{B}, \bar{A})$, but not both. But if it contains either $(\bar{B}, A)$ or $(\bar{B}, \bar{A})$ but not both, the required logical generality remains far short of attainment: as a consequence, Frege's basic sign gets inevitably devoid of the advantages that Frege finds in the negation of the 'non-logical' possibility in $(B, \bar{A})$ and, as a logically primitive sign, it becomes senseless. As a generalisation of function theory, then, his logic would turn up to be an extension of function theory in its special case when its basic notion is not trivial but nonsensical, hence, an extension of function theory in its special case when what becomes nonsensical is the theory as such.

1.6. *Frege's New Way of Conferring Generality: Empty Placeholders in the Context of the Conditional*

Returning to the issue of the putative use the letter '*x*', in Frege's notation, for expressing generality, consider once again the examples of Frege's notation above, with regard to which he notes that

> We must likewise distinguish the case of one concept being subordinate to another from that of a thing falling under a concept although the same form of words is used for both. The examples given above
>
> 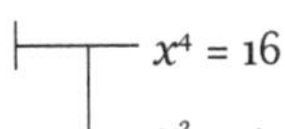
>
> and $\vdash 2^4 = 16$

> show the distinction in the concept-script. The generality in the judgement
>
> 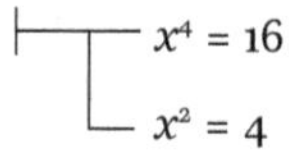
>
> 'All square roots of 4 are 4th roots of 16' is expressed by means of the letter x, in that the judgement is put forward as holding no matter what one understands by x.[24]

It would be expedient to observe that in the case of 'a thing falling under a concept' (presumably that of " 'the number 2' falling under the concept 'the square root of 16' " exemplified by $\vdash 2^4 = 16$) Frege quite manifestly avoids using 'letters', since in the case of doing so, as he states elsewhere in his posthumously published essays, the equation $x^4 = 16$

> asserts nothing, and so there admittedly can be no question of using the letter here to indicate—to confer generality on the thought. What we really have here is the designation of a concept, and the challenge to cite objects (in this case numbers) which fall under it. In that case the 'x' is used in the way I used the letter 'ξ' in the first volume: it occupies the argument-places, so that we can recognize them as such.[25]

What we can plainly see from the analysis of these two passages from Frege's writings is that whatever really forms the basis of his logical grammar, this can be neither the function/argument distinction as such, nor his variable/constant distinction, since the latter only follows, as a corollary, from the application of the former. It is clear that in the examples above the function/argument distinction, by itself, does not and cannot contribute to the expression of generality: what we have here in either of '$x^2 = 4$' or '$x^4 = 16$' taken alone is the letter 'x' in the role of Frege's letter 'ξ' marking no more than a mere gap within "the designation of a concept, and the challenge to cite objects (in this case numbers) which fall under it." And the only answer as to what it is that confers generality, if not the letter 'x' as such, is one that follows from Frege's own dictum: "It is enough if the proposition taken as a whole has a sense; it is this that confers on its parts also their content."[26] For, in the light of this dictum, the only sense 'generated' by the proposition in

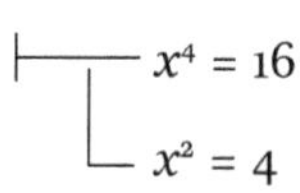

---

[24] *Ibid.*

[25] G. Frege, 'Logical Defects in Mathematics', *Posthumous Writings*, p. 163.

[26] G. Frege, *Die Grundlagen der Arithmetik: eine logisch-mathematische Untersuchungen uber den Begriff der Zahl*, X,71; cf. also G. Baker and P. M. S. Hacker, *Frege: Logical Excavations*, p. 206 (note 8).

as a whole happens to be exclusively that of the conditional: what Frege's assertion stroke means here is that the conditional in

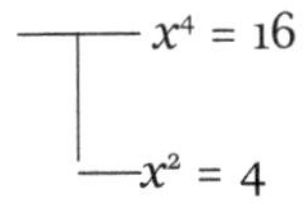

is true. However, this by itself gives no warrant whatsoever that either both or at least one of '$x^2 = 4$' or '$x^4 = 16$' can also be true. What it gives warrant to, rather, is the possibility that either one or both of them may be false. But the latter possibility (both of them being false) arises even if the letter 'x' does not take any number as a value. For the truth of the conditional obtains, among other things, due to the falsehood of both '$x^2 = 4$' and '$x^4 = 16$', and the mutual falsehood of both $x^2 = 4$ and $x^4 = 16$ (hence, the truth of the whole of the conditional) obtains, in turn, even if nothing is substituted for 'x'—leaving an empty space intact leads to the same effect. Moreover, in the light of Frege's own recognition that "the numerals are absolutely *not* proper names or objects ... so that a sentence like 'four is a square number' does not express the subsumption of an object under a concept"[27] this possibility becomes sheer necessity, precisely in the numerical examples of the logical conditional above that he uses to substantiate his view. In this way, the conditional appears as holding universally quite regardless of or, rather, precisely contrary to Frege's putative principle of compositionality: the reference of the whole of the conditional (its truth value) holds regardless of its being the function of the referents of its constituent parts, for the latter either may or else may not have referents as such. The conditional thus appears to confer generality on its constituent parts quite regardless of whether the letter 'x' is used in these constituent parts to indicate and, thus, to 'confer generality on the thought', or it is used as the letter 'ξ' whose role is merely to occupy the argument-places so that we could recognize them as such. And, if so, the letter 'x' in the meaning of Frege's letter 'ξ' will do as well, let alone the fact that in Frege's showcase conditional containing numerical expressions as its parts, the letter 'x' in the role of 'conferring generality on the thought' turns out to be demonstrably dispensable. Thus, contrary to a widely held view, the generality that Frege strives to attain in his *Begriffsschrift* arises, in fact, not due to Frege's use of 'letters' as *variable signs* as such but solely due to the use of Frege's *constant sign* of the conditional, which Frege

[27] G. Frege, *Nachgelassene Schriften*, p. 282.

chooses as the basic sign owing to its indeterminateness and, thus, generality of its logical meaning for the expression of logical relations as such.[28] The sign of the conditional in this way becomes the basic 'form-word' of Frege's universal logical grammar, conveying the basic structural meaning of sentence-structure in his 'language of pure thought'—quite regardless of his use of variables, neither as morphological constituents of his lexicon (Frege's use of *roman* letters), nor directly in the syntactic component (Frege's use of *gothic* letters) with the use of the concavity as their placeholder (as will be shown in a sequel). And, as such, it can be shown to be performing not only the same basic structural function as Chomsky's sentence formation rule[29] but also as necessarily producing the same basic 'semantic' effect as in the sentence obtained by the application of Chomsky's sentence formation rule in "Colorless green ideas sleep furiously."[30]

1.7. *Schröder's Objection Revisited*

As has been shown above, the generality that Frege attains in the logical grammar of his *Begriffsschrift* happens to be based on the generality of his logical sign for the conditional, quite despite Frege's attempts to present the function/argument distinction as the real source of generality being conferred on the thought by the use of variables in the argument places of his functional expressions. A similar observation to the same effect was made by Schröder in his review of Frege's *Begriffsschrift*: "The conditional wording, therefore, misleads one into the unintended interpretation of the antecedent as holding universally—into an imputation of "generality"—about which, by the way, the author later[31] makes some very pertinent remarks."[32] And even though what Schröder calls 'an imputation of "generality"' is, in his view, no more than just a misleading impression arising from the unfortunate wording that Frege uses, the real issue is not that trouble-free and cannot be discounted so easily with a hasty dismissive remark like Schröder's. This issue requires a more thoroughgoing

[28] Cf. Frege's elucidations as to the generality of the logical meaning of his sign for the conditional in *Boole's logical Formula-language and my Concept-script*, pp. 49–50.

[29] Chomsky, N., *Syntactic Structures*, p. 26.

[30] *Ibid.*, p. 15.

[31] *Begriffsschrift*, p. 19.

[32] Review of Frege's *Conceptual Notation* by E. Schröder, in Gottlob Frege, *Conceptual Notation and related Articles*, p. 227.

analysis that will show that there is much more to what he misguidedly spots as the result of Frege's "misleading wording."

Rather, it seems to be Schröder himself who happened to have been misled by Frege's 'pertinent remarks', namely by his introduction on p. 19 of his *Begriffsschrift* of his inscription of the gothic letter:

$$\text{——}\underset{\smile}{\mathfrak{a}}\text{——}$$

placed in a concavity in the middle of the content stroke that he uses as a designation for the (possible) logical content of the functional sign to which this stroke is prefixed:

$$\text{——}\underset{\smile}{\mathfrak{a}}\text{——}\ F(\mathfrak{a})$$

so as to be able to expose the relations of its possible content with that of the other signs (also functional expressions) appearing in a chain of proof, and thus dividing this logical content into two parts—one (to the left of the concavity):

$$[\text{——}]\,\underset{\smile}{\mathfrak{a}}\text{——}\ F(\mathfrak{a})$$

designating the general, and the other (to the right of it):

$$\text{——}\underset{\smile}{\mathfrak{a}}\,[\text{——}]\ F(\mathfrak{a})$$

the particular logical content of the functional expression in question. Quite regardless of whatever impression these remarks happened to have produced on Schröder, the imputation of generality by means of the conditional was, however, exactly part of Frege's strategy since, as he himself states in a piece of his posthumous writings (which Schröder had thus never read) with regard to the use of the letter x in his example of the quadratic equation in $x^2 - 4x + 3 = 0$:

> Regarded as a problem, the equation asserts nothing, and so there admittedly can be no question of using the letter x here to indicate—to confer generality on the thought. What we really have here is the designation of a concept, and the challenge to cite objects (in this case numbers) which fall under it. In that case the 'x' is used in the way I used the letter 'ξ' in the first volume: it occupies the argument-places, so that we can recognize them as such. Once you have solved the problem, you may assert the sentence

> 'If $x = 1$ or $x = 3$, then $x^2 - 4x + 3 = 0$', and here the letter 'x' is once more used to indicate as above, to make the thought general. Whatever numerical sign may be substituted for 'x', we always obtain a true sentence, either because the condition is not fulfilled, in which case it is all one whether the consequent is true or false (for instance if you substitute '2'), or because the condition is fulfilled, in which case the consequent is then also true.[33]

The generality allegedly attributed to the 'indicating' use of the variable in $x^2 - 4x + 3 = 0$ depends here specifically on Frege's use of the conditional or, to use Schröder's turn of phrase—on 'the antecedent as holding universally'; for regarded not as a problem but as part of the conditional and thus according to Frege's context principle, the equation in $x^2 - 4x\ 3 = 0$ already asserts, but what it then asserts is either true or false depending upon *what* we substitute as an argument for x and so it does not appear as holding generally, while it is exactly the conditional that turns up to be valid *whatever* is substituted as an argument for x. It would be instructive, then, to compare it with Frege's elucidation of another example of a quadratic equation (which happens to be the same quadratic equation but in its more general form), in which he speaks not about logical but about mathematical functions:

> When we write:
>
> $x^2 - 4x = x(x - 4)$,
>
> then we have not set one function equal to another, but we have only set the values of the functions equal. And when we understand this equation as being valid whatever is substituted as an argument for x, then we have expressed the generalization of an equation.[34]

But even though "in '$x^2 - 4x = x(x - 4)$,' the left side considered in isolation indicates a number only indefinitely, and the same is true of the right side,"—and, thus, regardless of the use of the variables to indicate—generality, as Frege explains, is "something that is not contained *in the left side* by itself, nor in the *right side*, nor in the 'equals' sign;" for we express it only "if we *combine the two sides* to form an equation," with "the same letter for both sides."[35] Thus the use of the variable appears to be only a necessary, but not a *sufficient* condition for the expression of generality, for it is only when we "choose the same letter for both sides" and, by doing so, we also "set the values of the functions equal" that we obtain the

[33] G. Frege, 'Logical Defects in Mathematics', *Posthumous Writings*, p. 163.
[34] G. Frege, *Funktion und Begriff*, p. 9.
[35] *Ibid.*, p. 11.

generalisation of an equation, i.e., understand this equation as being valid whatever is substituted as an argument for x.

Notably, what Frege speaks about is, at root, functions and *their values*, for the key phrase in his elucidation is that in $x^2 - 4x = x(x - 4)$, "we have not set one function equal to another, but we have only set the *values of the functions* equal." So it is clear, then, that the ultimate, and thus real, source of generalisation in question springs out of functional generalisation as such where, as Frege reminds, "just as we indicate a number indefinitely by a letter, in order to express generality, we also need letters to indicate a function indefinitely,"[36] as in the general case of functional representation as such that we find in $f(x) = y$. Owing to the generality of the sign of the function in this expression, this functional sign is a common designation for the functions to the left and to the right of the sign of equality. But it would be misleading to rewrite the equation as $f(x) = f(x)$ for if we take, along with Frege, the equals sign as a sign for identity, it would produce the wrong impression that functions are identical, but indeed they are not. What '$f(x) = f(x)$' designates is function as type and what we have in $x^2 - 4x = x(x - 4)$ are necessarily *instances* of function, but not function as type *per se*. Of course, as instances they, too, may be identical and thus conform to the pattern in '$f(x) = f(x)$', but this would be a limiting case only. The essence of the equation (and thus of the generality obtained in it) lies in something that is *not* inscribed in $f(x) = f(x)$, but appears only in the formula for functional generalisation, $f(x) = y$, namely in another, so-called dependent, variable y designating another variable though correlated value—the numerical value of the function itself as opposed to the value of its argument. Why we obtain generality is exactly because the values that we set as equal are—as is seen from Frege's explanation above—precisely the values of this dependent variable appearing, even if not overtly, on both sides of the sign of equality, and the sign of which, by setting its values as equal, we also make coreferential in both occurrences in the same way as we do it for the independent variable x by simply inscribing both its occurrences (to the left and to the right of the 'equals'-sign) by the same letter which thus "*retains* in a given context the meaning once given to it."[37]

Thus, as Frege makes absolutely clear in his explanations cited above, it is not only the generalising effect arising from the use of the independent

---

[36] *Ibid.*

[37] *Begriffsschrift*, § 1.

variable x to indicate numbers indefinitely, and not even the respective use of the dependent variable y to which Frege definitely refers by saying that "in '$x^2 - 4x = x(x - 4)$' *the left side* considered in isolation *indicates a number only indefinitely*, and the same is true of *the right side*" (taken as a whole), that makes the expression of generality attainable. And—what is essential—not only in this particular example. For, as Frege himself concludes, "Admittedly what we express is the generality of an equality; but primarily it is a generality."[38] To express what Frege refers to as 'a generality', the use of either of them will be seen as utterly insufficient. For the attainment of generality this would make up only a necessary condition, but evidently not a sufficient one. And what makes up the sufficient condition is not very hard to guess: it is a *simultaneous* use of both of these variables, without which the very notion of a function simply loses sense.

1.8. *Frege's Hidden Agenda*

As we can now see, the expression of the generalisation of an equality and, thus, the expression of what primarily is a generality makes sense only in the presence of at least two variables functionally related to one another on *either* side of the sign of equality, i.e., it necessarily requires that the values of the functions (and not only of their arguments) should, too, be *indefinite*. Notice, however, that if we fix a *definite* value of the function and, in doing so, replace the dependent variable by a constant sign, any talk about generality becomes void. In order to see why it is so, consider a function as a relation between the elements of its domain (which are the values of x as an *independent* variable in $f(x) = y$) and its range (which are the values of y as a *dependent* variable). Thus, if x is an element of the function's domain, then $f(x)$ is the element of the function's range that *corresponds* to it, so we have an example of the functional mapping (e.g. $x^4 = y$) as below:

Elements of domain x	Function f	Elements of range $f(x)$
1	(1, 1)	1
2	(2, 16)	16
3	(3, 81)	81
4	(4, 256)	256

$f(2) = 16$

[38] *Ibid.*

Now, if we pick one of the elements of the range (as in the table above), then what we have is only an *instance* of the functional relationship and, thus, not a functional generalisation as such. Moreover, this instance of the functional relationship would also necessarily involve a certain value of the independent variable, so that the whole of our functional expression would become a constant sign. And if we now insert the inscription of a variable in the place of the now constant value of the argument—which is exactly what Frege does converting an example like '$x^4 = 16$' into a logical function[39]—this will not change anything: it would yet be only *one instance* of functional mapping in '$x^4 = y$', and in this instance the generality that we were once so happy with would have been irrevocably lost. For, regardless of whether we conceive the value for the argument as a proper name (as in $f(2) = 16$) or whether we leave it in disguise as a *pseudo*-variable (as in $f(x) = 16$), what we inexorably get will in either case be nothing but a *particular* case of functional generalisation in "$f(x) = y$"—and never anything more than just that:

Elements of domain x	Function f	Elements of range $f(x)$	
✗ 1	(1, 1)	✗ 1	
$x = 2$	(2, 16)	→ 16	$f(2) = 16$
✗ 3	(3, 81)	✗ 81	
✗ 4	(4, 256)	✗ 256	

By this account, any attempt to upgrade this particular case to the status of functional *generalisation* would involve a fatal error. Not even if we call it a generalisation of a function of a different type. Moreover, since we can observe that what Frege puts in the context of (as we have seen above) a *necessarily constant* basic sign of his logical grammar—as a necessary precondition of becoming part of this context—it should also *necessarily be constant*[40] or otherwise it would be rejected by this context,[41] we can also predict this will hold for the argument-sign in the functional generalisation in $f(x) = y$ as well. But the only case when x could be regarded

[39] ... i.e. a function whose value is always a truth-value.

[40] ... for it is exactly taking a particular case of the functional generalisation in $f(x) = y$, hence, the case when it necessarily acquires the character of a constant sign, which can only form the basis for the introduction, by Frege, of his notion of *logical* function.

[41] ... exactly this also happens to be the source of what Westerståhl (1986:118) observes as the falsehood of $\forall x\Psi(x)$ when functions inscribed as $\Psi(x)$ are general.

as constant is when it is used in the role of Frege's letter 'ξ', "and so there admittedly can be no question of using the letter x here to indicate—to confer generality on the thought."

Thus, the representation of the letter x in the mode of its being a type cannot be mixed with the representation of the letter x when its designation is that of a token, for the *referents of type and of token are essentially different*. But in Frege's system, as we can see from above, *this difference is completely obliterated*; moreover, he puts it forward as his basic principle in Basic Law V of the *Grundgesetze*, which he upgrades to the status of a basic logical truth. With regard to this Sluga, however, notes:

> If '$\dot{x}f(x)$' stands for the value-range of the function $f(x)$, we can write that law informally as
>
> $$(1) \quad \forall x(fx = gx) \leftrightarrow (\dot{x}fx = \dot{x}gx)$$
>
> The question remains why Frege should consider this law a basic logical truth. It is not immediately clear how this claim could be argued. On the contrary, it might look more plausible to deny that (1) could possibly be a logical truth since the sentence on its left side makes a statement about functions and the sentence on the right side is about value-ranges. Frege insists on a sharp distinction between functions and objects, so it might be argued that the one side of the equivalence is made true by a quite different state of affairs from what makes the other side true. Even if (1) is true, there is then no reason to think that it is logically true.[42]

But, as we have just seen, there precisely *is* reason to think that it is exactly to the contrary. There is more going on in Frege's function-manipulation that meets the eye, and certainly more than meets a naïve contentment with his 'explaining' variables by mere analogy with the use of pronouns 'to indicate indefinitely'. The question that now naturally arises is whether Frege really fulfils the task that he claims to have fulfilled. For what he claims is that he has rebuilt generality by placing particular instances of his mathematical functions in the context of his conditional where, as he maintains, "the letter 'x' is once more used to indicate as [in the context of a mathematical equality] above, to make the thought general."[43] But this happens to be only a misleading impression, for if he (or whoever else after Frege) thinks that in his new context of what Schröder calls 'conditional wording' it is also true that "[t]he sentence now says—and just this is what constitutes its generality—that a true thought is expressed in

[42] H. Sluga, *Gottlob Frege*, p. 149.

[43] 'Logical Defects in Mathematics', *Posthumous Writings*, p. 163.

this way no matter what proper names of numbers may be substituted,"[44] then he appears to be grossly mistaken. For then he may be shown to have fallen victim to a great illusion.

To see how great is its extent, we will consider a simple expression in $\exists x(Fx \rightarrow Fx)$—known in modern logical parlour as the 'axiom of existence'—postulating the existence of objects, in our case numbers, whose proper names are assumed to be substitution values. But before doing that consider first an example of what Frege calls an existential judgement in

(8.1) 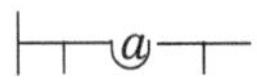 $a^2 = 4$

where an expression of 'the formula language of arithmetic', namely

(8.2) $a^2 = 4$

is attached to Frege's expression of generality in the form of a concavity in which he places the letter '*a*' flanked on both sides by the sign of negation. Notice that it is not the use of the quantifier prefix containing Frege's equivalent of the modern existential quantifier that asserts the existence of some value for *a*. For, regardless of whether we use the quantifier or not, its existence already follows directly from (8.2) itself in virtue of a) the very nature of the function expressed by its generalisation in $f(x) = y$ and b) the fact that $a^2 = 4$ presents itself as a particular case of this functional generalisation due to a fixed value of y in $a^2 = 4$. Since as a means for the expression of an existential statement this quantifier prefix is absolutely redundant, the same existential statement happens having already been asserted by $a^2 = 4$ itself, Frege evidently has an agenda here. The covert, underlying sense of what Frege does is then something different from what he says overtly: what Frege really wants to show, by this very example, is obviously much more general than the possibility of an obviously redundant restatement of the existence of some value for *a*. For, evidently, his task is not only the illustration of the expression of the various forms of judgement by his form of notation, of which he *speaks* and to which the attention of the reader is thus immediately attracted, but also the demonstration that the formulas that he at the same time *shows* are putative (I dare not say feigned) part of his concept-script. And, if this tactic succeeds, then one may falsely infer from

[44] *Ibid.*, p. 162.

this that the strategy of functional generalisation that he employs is not based on a fallacy, but allegedly on a sound understanding of basic concepts from within arithmetic, reportedly "based on principles which 'do naturally and necessarily belong to the Philosophy of letters and speech in the General', as distinct from 'instituted' grammar . . . peculiar to any one Language'."[45] For, principally, his logical signs may be shown to be directly applicable to the formula language of arithmetic in one case only, when the functional expression appearing in his definition of the quantifier

(8.3) $\vdash\!\!\overset{\frown}{a}\!\!- F(a)$

could itself be seen as a sort of functional generalisation over both $F(x) = y$ (where y is universal) and $F(x) = n$ (where n is some particular value from the range of the values for y). Because this would be a necessary condition for the applicability of the quantifier not only directly to the functional generalisation in $F(x) = y$ (as in the case of $\forall x(x = x)$), but also indirectly (by the use of the dual of Frege's quantifier in $\neg\forall x\neg$) to its particular instances in $F(x) = n$, so that both of $F(x) = y$ and $F(x) = n$ would then be regarded, on equal grounds, as being themselves nothing more but particular instances of his truth-functional generalisation in $F(a)$ that he attaches. It is exactly this truth-functional generalisation that, due to its very capacity of being a functional generalisation (even if it were one of the now already two parallel—hence mutually competing and thus either of them excluding the other one—functional generalisations) would indeed revert the no-matter-what-we-substitute-for status of the letter x.

Notice now that in one of the particular cases of the new functional generalisation (of all functions as instances of truth-functions)—the former functional generalisation in $F(x) = y$ now supplemented by another functional expression in the place of y (and thus subject to universal quantification), the imputed role of the letter x as a means for conferring generality to thought appears to be itself the product of, or have its source in, the already noted generalising character of the values of functions in $\Phi x = \Psi x$,[46] thus, not in Frege's truth-functional generalisation as such. And, in the other of the particular cases in question, the putative

[45] Roy Harris and Talbot J. Taylor, *The Western Tradition from Socrates to Saussure*, p. 119.

[46] . . . where Φ and Ψ may eventually turn up as the same function, then the generality arising from the equality of its own values, $y = y$, may be rewritten as the familiar $x = x$.

use of the letter to confer generality seems only to be supported by the mere presence of the same variable in the concavity, despite the tension created by the apparent singularity of the same letter in the expression of the function. To resolve this tension Frege introduces his notation for existential judgements like "there is at least one square root of 4," which Frege depicts as the negation of the generalisation of the negation of the equation, i.e., $a^2 = 4$, in

(8.1) 

and $F(a)$ in

(8.4) 

in which the presence not of Frege's quantifier but of its dual becomes mandatory.

From inspection of (8.1) and (8.4) we can see that Frege clearly depicts $a^2 = 4$ as intersubstitutable with $F(a)$. At first sight, then, (8.4) presents itself as a self-evident generalisation of (8.1). Notice, however, that it presents itself as *self-evident only seemingly*. What is obvious—though not *immediately* self-evident—is that for $F(a)$ to appear as a function of the same kind as $a^2 = 4$, i.e. as a function from individuals to truth-values, it has necessarily to be in the context of the conditional, since otherwise the parallel generalisation in (8.3) would be false, which also means that (8.3) would be false as a *generalisation*—the fact already noted by Westerståhl.[47] So, for $F(a)$ to remain a function of the same kind as $a^2 = 4$, it can only appear either in the form of the already mentioned 'axiom of existence', namely

(8.5) 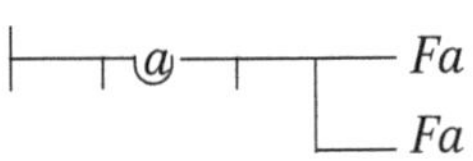

or, in virtue of Frege's principle of structural equivalence of particular and existential judgements,[48] as

(8.6) 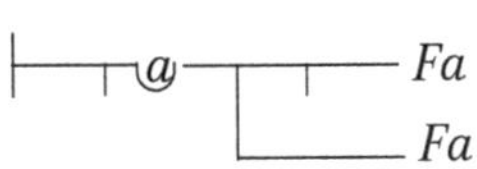

[47] See footnote 41.

[48] *Boole's logical Calculus and the Concept-script*, pp. 20–21.

or as a particular case of (8.6) in

(8.61) 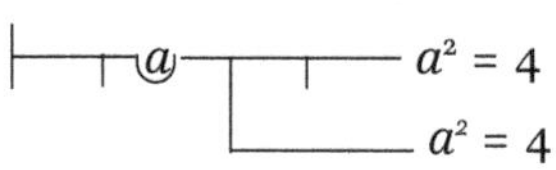

On a more careful inspection of both (8.6) and (8.61) it will be seen that the required status of the gothic letter as an expression of generality is unattainable, not even in the context of the conditional. For it happens to be exactly the context in which Frege's claim is demonstrably false.

1.9. *The Fregean Quantifier and the Philosophical Clarification of Generality: Frege's Misjudgment and Heidegger's Prophecy*

For the sake of expediency, consider Frege's explanation of generality in algebraic expressions:

> *In the familiar sentence*
>
> (9.1) $(a + b) \cdot c = a \cdot c + b \cdot c$
>
> we have the ordinary use of letters to indicate. They serve here to confer generality on the thought. They stand in the place of proper names, but are not such (pronouns). You always obtain a particular case of the general sentence when you substitute the same proper name of a number throughout for the letter *a*, and similarly for *b* and *c*. The sentence now says—and just this is what constitutes its generality—that a true thought is expressed in this way no matter what proper names of numbers may be substituted.[49]

Hence, what constitutes the generality of (9.1) is nothing other than the logical value True acquired by the (possibly) infinite conjunction of its particular cases each of which is also true no matter what numbers we *substitute* for argument values. In other words, the generality that he exposes in (9.1) is *logical* generality expressed as the validity of the propositional formula

(9.11) $p_1 \wedge p_2 \wedge p_3 \wedge \ldots \wedge p_k \wedge \ldots \wedge p_n$

To falsify this formula one needs to find at least one case in which p would be not true, i.e. one needs such a proper name of a number, *the substitu-*

[49] *Logical Defects in Mathematics*, p. 162.

tion of which as a value for the argument produces the value False. Note that the falsehood of this particular case is not yet the failure of the generality of (9.1) which is the truth of (9.11) as a whole. The falsehood of any particular case would be the falsehood of that particular case only. This falsehood is only the starting point and it yet has a job to do. For the failure of generality comes only as a value False of the *whole* of the conjunction in (9.11), hence, as falsehood of a propositional connective as such, and the value of p affects the value of (9.11) in this way and not the other only in virtue of the fact that this propositional connective is a conjunction. In other words: whatever value for the value of the argument we choose, to say that (9.1) holds generally is the same as to assert the truth of the infinite conjunction of p's in (9.11). And this in turn derives its value from the truth-functional values of each p. Notice, however, that the function about which we say that it holds generally is not a truth-function taking us from *individuals* to truth-values, for it has values True and False not only for the values of the function itself, but for its arguments as well.

Consider now what the conditions for the expressions in (8.6) and (8.61) with a gothic letter a as holding generally would be, in the same manner as we did it with regard to (9.1). The expression corresponding to (9.11), with the negation to the left of the gothic letter being (temporarily) omitted, would then take the form:

$$(9.2)\quad (p_1 \rightarrow\sim p_1) \wedge (p_2 \rightarrow\sim p_2) \wedge (p_3 \rightarrow\sim p_3)\wedge\ldots\wedge(p_k \rightarrow\sim p_k)\wedge\ldots\wedge(p_n \rightarrow\sim p_n)$$

or, in a corresponding functional representation,

$$(9.21)\quad (F_{(1)} \rightarrow\sim F_{(1)}) \wedge (F_{(2)} \rightarrow\sim F_{(2)}) \wedge (F_{(3)} \rightarrow\sim F_{(3)})\wedge\ldots\wedge(F_{(n)} \rightarrow\sim F_{(n)})$$

Since the negation to the left of the concavity would be the negation of generality, it would then be the negation of the falsehood of the conjunction in (9.21), in much the same way as it would be the negation of the truth of the conjunction in (9.11). And what we would then naturally expect, since (infinite) conjunction is generalisation, is that this generalisation would be falsified (and thus, as a false generalisation, will be made true) by the occurrence of at least one specific value of the argument in one of the p's in (9.2)—and, respectively, one of the F's in (9.21)—such that it makes this p alias F false. But if one looks at the truth-functional values of the instances of the conditional in (9.21) represented in Fig. I:

$$\overbrace{(\underbrace{F_{(1)}}_{\text{False}} \rightarrow\sim \underbrace{F_{(1)}}_{\text{True}})}^{\text{True}} \wedge \overbrace{(\underbrace{F_{(2)}}_{\text{True}} \rightarrow\sim \underbrace{F_{(2)}}_{\text{False}})}^{\text{False}} \wedge \overbrace{(\underbrace{F_{(3)}}_{\text{False}} \rightarrow\sim \underbrace{F_{(3)}}_{\text{True}})}^{\text{True}} \wedge \ldots \wedge \overbrace{(\underbrace{F_{(n)}}_{\text{False}} \rightarrow\sim \underbrace{F_{(n)}}_{\text{True}})}^{\text{True}}$$

it will immediately be seen that finding some instance of $F(a) \rightarrow\sim F(a)$ that satisfies this condition will not suffice. For, instead of negating the falsehood of the generality in (9.21), this in fact will only confirm it. Moreover, it will be evident that the only way for the falsification of the false generality in (9.21) to make sense at all is not to make $F(a)$ true under all substitutions (for this only brings us to the falsehood of all instances of $F(a) \rightarrow\sim F(a)$ and, so to speak, aggravates the falsehood of the whole conjunction) but, on the contrary, to make $F(a)$ under all substitutions false (so as to make the rest of the instances of $F(a) \rightarrow\sim F(a)$ true as well). Thus, Frege's claim appears to be demonstratively false, since the only option in which the (infinite) conjunction in (9.21)[50] turns out to be true is when all the argument places in the functional expressions in (9.21) are *not* filled, for only this will result in the truth of the consequent in each instance of $F(a) \rightarrow\sim F(a)$ and, thus, in the truth of each instance of $F(a) \rightarrow\sim F(a)$ itself.[51] But this can happen if and only if the letter a appears in both (8.6) and (8.61) only *in the role of Frege's letter* ξ that merely "occupies the argument-places, so that we can recognize them as such," but *not as an expression of generality*. Moreover, it is not only that there is no question of using this letter here to indicate—to confer generality on the thought. For what clearly is also not the issue here is the role of the letter ξ as Frege's correlate of the letter x in an expression of a logical function, namely in a function from objects to truth-values. For, as Frege points out speaking about his use of the letter x in the role of the letter ξ: "What we really have here is the designation of a concept, and the challenge to cite objects (in this case numbers) which fall under it."[52] But, as we have just seen, in the case of $\Phi(a)$ understood as a *logical* function there is not much of a question of whatever objects (numbers including) falling under

[50] . . . along with the respective formulas in (8.6–8.61).

[51] . . . note that what we are dealing here are *substitution* instances of an open formula, and not the open formula as such.

[52] *Logical Defects in Mathematics*, p. 163.

Frege's designation of a concept; hence, there is not much of a question of Frege's '$\Phi(\ )$' as the designation of a concept, either.

What this clearly demonstrates is, in effect, that Frege's argument that the logical generality of the conditional, owing to which he chooses it for his basic logical sign, can be effectively augmented by the generality of functional expressions, turns out to be based on nothing but delusion. Thus the very idea of the possibility of the incorporation of the constant/variable distinction of function theory *directly* into his signs of logical relations turns up to be demonstrably false. For it can be seen that it is the nature of the basic logical relation which effectively precludes the use of Frege's variables as a means for expressing *logical* generality. And this shows how far short of attainment Frege's technical expression of generality brings him, if it is meant as his way of attaining the universal, and through it also philosophical, status of his *Begriffsschrift*. For the foregoing analysis reveals that what in Schröder's case was really misleading was not the "interpretation of the antecedent as holding universally" that "misleads one ... into an imputation of 'generality'," but Frege's "pertinent remarks"[53] about his technical expression of generality that misled him into an impression of its being itself a *philosophical clarification* of generality. And, as a corollary, what this also shows is how far short of attainment Frege's device of quantification brings us, if it is meant as our way of attaining to the universal, and thus also philosophical, status of symbolic logic.

The failure of Frege's strategic plan to combine his basic logical sign with his functional representation of conceptual structure can be elucidated somewhat further if we take a closer look at the logical properties of the so-called axiom of existence in

(9.3) $\exists x(Fx \rightarrow Fx)$

Owing to the presence of the existential quantifier customarily paraphrased as "there is at least one x such that ...", (9.3) is widely viewed as stating the existence of some x such that it has a certain property, F. In the axiom of existence this fact seems to get further support from the status of (9.3) as an axiom and, thus, as an analytic expression which, as an analytic expression, should be necessarily true, i.e., true under all possible substitutions and, therefore, no matter what we choose for the value of x.

[53] *Begriffsschrift*, §11.

Consider, however, the representation of this axiom in Frege's symbolism, where (9.3) assumes its respective form in:

(9.31) 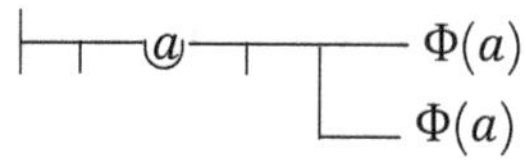

It will immediately be seen, on a closer inspection of (9.31), that the part of the formula to the right of Frege's sign of generality expresses nothing more and nothing less but the negation of the principle of *tertium non datur*. This becomes fairly evident since the content-stroke directly flanking the concavity is the content-stroke of the conditional in much the same way as the content-stroke to the right of the concavity occurring in front of $\Phi(a)$ is the content-stroke of $\Phi(a)$. Therefore, what this content-stroke with the sign for negation attached to it can only express is the assertion of the fact which the conditional itself negates, i.e. the statement to the effect that the antecedent is true while the consequent, false. Then, owing to the equivalence of $P \rightarrow Q$ and $\sim P \vee Q$, i.e. the equivalence which is used in *Principia Mathematica* for the definition of the Philonian conditional, which in our case is the equivalence of $P \rightarrow P$ and $\sim P \vee P$, the function of the negation in front of the conditional in (9.31) is obviously that of invalidating the principle of *tertium non datur*.

Since, according to Frege, "the stroke occurring to the right of the concavity is the content-stroke of $\Phi(a)$—we must here imagine something definite substituted for a,"[54] the substitution of something definite for yields

(9.311) 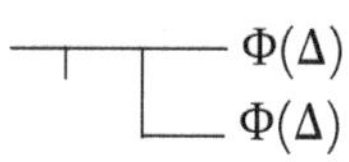

in which functional expressions acquire definite truth-values, hence, can be replaced by respective propositional variables and rewritten, in a more conventional manner, as

(9.312) $\sim(p \rightarrow p)$

Then, owing to the fact that (9.312) is invalidating the principle of *tertium non datur*, the presence of the additional negation appearing to the left of the concavity can only be seen as indicating that the negation that we are

[54] *Begriffsschrift*, §11.

dealing with in this case is not strong (contradictory) but weak (contrary) negation. But the weak character of the negation involved points out to the fact that the general validity of (9.31) obtains only in virtue of its being truth-functionally equivalent with

(9.32) $p \vee (\sim p) \vee [\sim (\sim p)]$,

due to the presence of a third, closer unidentified truth-value of *p*. And this leads to a disclosure of a very interesting truth-functional property of the conditional in (9.31). For, if we consider truth-table definitions of propositional connectives in Fig. II:[55]

A	B				$B \to A$
p	$\sim p$	$\sim (\sim p)$	$p \vee (\sim p)$	the negation of $p \vee (\sim p)$	
1	½	0	1	½	½
½	0	1	½	0	1
0	1	½	1	½	0

we can see that the use of the conditional in (9.31) conforms to the second line of the table, and it does so for at least three reasons. Firstly, it is the only line where the initial value of *p* accords with the status of the $\Phi(a)$ in (9.31) prior to any substitutions, for it is natural to think of $\Phi(a)$ as if it were an $\Phi(\Delta)$ in which the values of Δ remain unknown or inaccessible (for the reason of the infiniteness of the Δ's, for example), so that the value of $\Phi(a)$ may be either True or False. Secondly, it is the only line where the truth-value of $\sim P \vee P$ appears to be invalidating the principle of *tertium non datur*. Thirdly, Line 2 makes the only case when the negation of the negation of $\sim P \vee P$ leads to the truth-value True of the whole of (9.31). Incidentally, it happens to coincide with the truth-value of the double (weak) negation of *P* as such. And, most importantly, as is also seen from Fig. II, it is the only line in which the conditional happens to be valid.

[55] ... where *A* and *B* stand for the respective elements of Frege's conditional, and 1, ½, 0 are the designations of the truth values "True," "either True or False," and "False," respectively.

Then, in the case of B (which is the antecedent of the conditional) being False, $B \rightarrow A$ (the conditional itself) is True, i.e., this turns up to be indeed the only case when, "we always obtain a true sentence, because the condition is not fulfilled, in which case it is all one whether the consequent is true or false."[56] Notably, Frege's case of being 'all one whether the consequent is true or false' happens here to be exactly the truth-value ½, "either True or False," of the consequent. But this effectively precludes the alternative, which Frege thinks is also making the case, when he says that

> Whatever numerical sign may be substituted for 'x', we always obtain a true sentence, either because the condition is not fulfilled, in which case it is all one whether the consequent is true or false (for instance if you substitute '2'), or *because the condition is fulfilled,* in which case the consequent is then also true.[57]

Moreover, even though it happens to be the case of being "all one whether the consequent is true or false", it is obvious that in the axiom of existence we have only a limiting case of *the sentence being true even when the condition is not fulfilled,* because of the formal identity of the antecedent and the consequent, both of them being the same $\Phi(a)$. For, if $\Phi(a)$ *must* have the truth-functional value of $\sim p$ as an antecedent, then as a consequent it must have such a value of p which *has to be identical* with the value of $\sim p$, since otherwise it would have to be the result of the substitution of two different values for Δ in the occurrences of $\Phi(a)$ as an antecedent and as a consequent, and thus it will have to be the result of the violation of the condition of identity. Therefore, since the value of $\sim p$ is False, the value of p cannot have True as an ultimate instance of "either True or False" (in the sense of our not knowing or our lacking information as to its actually being True or False)—as such an instance, it can only have the value False. But this can only happen when we leave either of the occurrences of $\Phi(a)$ without any substitution of any definite value of Δ, since this is the only case when, in Frege's own terms, both of them can be false. And this results in the necessity of the appearance of a, in (9.31), only and exclusively in the role of Frege's letter 'ξ' which only "occupies the argument-places, so that we can recognize them as such."[58] But this is also the case when there can be no question of using this letter to indicate—to confer

---

[56] *Logical Defects in Mathematics*, p. 163.
[57] *Ibid.*
[58] *Ibid.*

generality on the thought. For this letter is nothing more than an abbreviation for an *empty* placeholder for the sign of the argument.

Thus, despite the fact that what Frege apparently deals with is *prima facie* a logic involving only two values, True and False, the logical reality of his quantifier-negation manipulations that he uses for the demonstration of the connection between the meanings of natural language items like *all* and *some* turns up, in effect, to be that of a trivalent logic. (Hence, the formal identity of the cross-quantifier equivalencies with the respective equivalencies involving modal operators, such as that of $\neg\forall x\neg F(x) \equiv \exists xF(x)$ with $\neg\Box\neg p \equiv \Diamond p$,[59] this formal identity being evidently conditioned by the plain fact that the sense of the modal operator "$\Diamond$" is essentially the same as that of an incremental truth-value "½" when p is interpreted as "either True or False.) But, as a corollary, Frege's quantifier expressions involving the conditional whose parts take over the role of the former subject and predicate turn up to be valid only on an empty domain. For the non-referring character of variables in these expressions implies that, for these expressions to be true, a domain of individuals should be empty. And, "to say that a supposed domain of individuals was empty could be at best only a roundabout way of maintaining that some complex of signs which had been supposed to express a propositional function did not in fact express anything at all."[60]

Furthermore, what this in effect demonstrates is that Frege's functional representation of conceptual structure (erroneously viewed as a representation of the structure of what is routinely referred to as an *atomic sentence*, even though, as we have already seen, the structure of a concept is radically different from that of a sentence as such) and the logical grammar of his *Begriffsschrift* are, even in principle, incompatible. For it could only be seen as if it were a compatibility of an expression containing Frege's letter 'ξ' in the antecedent and the functional expression of the form $f(x) = m$, in the consequent. The antecedent (with an empty placeholder for the sign of the argument) would then be necessarily false, the consequent, true or false (accordingly to what we substitute for x), and the whole of the conditional would then be true (being all one whether the consequent is true or false). But this is not possible for the simple reason that we would then violate the condition of identity. For we would

[59] Cf. also G. H. von Wright, *An Essay in Modal Logic*, p. 2; D. Føllesdal, *Referential Opacity and Modal Logic*, p. 12.

[60] W. Kneale, and M. Kneale, *The Development of Logic*, p. 707.

then have to assume that the $\Phi(a)$'s in question are identical, though in fact they are not.

The idea of compatibility would then stand or fall depending on whether the whole of the conditional turns up to be true or false. But the conditional cannot be valid because in the last analysis it turns up not to be consistent. For there can be no proof of its consistency, because for such a proof to become possible requires that the domain of individuals admissible as arguments to propositional functions should contain at least one such individual, whereas, as we have just seen, there can be no question of admissibility since the domain should, of necessity, be empty. But the attainment of this validity can be shown to be nothing else but the upshot of the whole of Frege's enterprise of constructing, for the first time, a truly philosophical grammar. This, however, happens to be exactly "the limits to the validity of a proposition" that Frege claimed to have been established by him for the first time.[61] But these appear to be the limits not in the positive sense assumed by Frege, but in the sense assumed by Heidegger when he spoke about "the barrier in the application of mathematical symbols and concepts (above all the concept of *function*)," and stated that "mathematics and mathematical treatment of logical problems reach limits where their concepts and methods fail; that is precisely where the conditions of their possibility are located."[62] Heidegger, however, did not provide any closer explanations as to what exactly the nature of this barrier might be. The main force of his argument he directed against the rising philosophical scope and claim of symbolic logic, most vivid in the late twenties and early thirties due to the work of Carnap. So Heidegger's argument was, of necessity, of a more speculative than strictly exegetic nature. This can be clearly seen in the following passage from his lecture course "Fundamental Questions of Metaphysics," of 1935–36, where he says of symbolic logic:

> There is an attempt here at calculating the system of propositional connections by means of mathematical methods; hence this kind of logic is also called "mathematical logic." It sets itself a possible and valid task. However, what symbolic logic furnishes is anything but a logic, i.e. a contemplation of the λογοσ. Mathematical logic is not even a logic of mathematics in the sense that it determines or could at all determine the nature of mathematical thinking and mathematical truth. Rather, symbolic logic is itself a type of mathematics applied to sentences and sentential forms. Every mathematical

[61] *Die Grundlagen der Arithmetik*, p. 1.

[62] M. Heidegger, *Neuere Forschungen über Logik*, col. 570.

> and symbolic logic places itself outside whatever realm of logic because for its very own purposes it must posit the λογοσ, the proposition, as a mere connection of concepts which is basically inadequate. The presumption of symbolic logic of constituting the scientific logic of all sciences collapses as soon as the conditional and unreflective nature of its basic premise becomes apparent.[63]

1.10. *GTS as Games with Tainted Strategies*

To see why the positing of the proposition as a mere connection of concepts is basically inadequate and what is wrong with the application of the concept of function to the solution of the living problems of the meaning of propositions, consider the basic premise of quantificational logic concerning the nature of quantifiers, as expressed by Hintikka:

> If the idea of quantifiers as higher-order predicates is right, then a first-order existential quantifier prefixed to an open formula says merely that the (usually complex) predicate defined by that open formula is not empty.[64]

Here, the idea of quantifiers as higher-order predicates is exactly Frege's idea of a first-order concept expressed by the (usually complex) predicate *falling within* a higher order concept (that of a quantifier), the latter saying, roughly, that the first-order concept in question has instances. The crucial difference that appears to be completely obliterated here by Hintikka's use of brackets is one between an *atomic* open formula, defining a predicate which is not complex, and a *non-atomic* open formula, defining a *usually complex* predicate. The source of this confusion and also the reason why the barrier in the application of the concept of *function* indicated by Heidegger have so far been left without due recognition may be seen in the general misunderstanding of the exact nature of Frege's replacement of the analysis of propositions into subject and predicate by his distinction between function and argument. To put it roughly: in the widespread *confusion* of the analysis of propositions into subject and predicate *with* his distinction between function and argument—the confusion that led to the misleading locution "predicate logic" instead of a more proper way of saying the same thing by using the more adequate term "functional logic." For what Frege in fact did was not a straightforward elimination

---

[63] M. Heidegger, *Die Frage nach dem Ding*, p. 122.

[64] J. Hintikka, *The Principles of Mathematics Revisited*, p. 69. Cf. also J. Hintikka, and G. Sandu, "Uses and misuses of Frege's ideas," *The Monist* 77, p. 283.

of the subject/predicate distinction in favour of his distinction between function and argument. As we see from his example in (5.1) which we reproduce here as (10.1):

(10.1) $$\vdash\!\!\!\begin{array}{l} \top\; x^4 = 16 \\ \llcorner\; x^2 = 4 \end{array} \qquad \text{and} \qquad \vdash 2^4 = 16$$

the subject/predicate distinction which Frege renders by means of his conditional to the left of "and" is not replaced by one of the possible ways of functional decomposition of $\vdash 2^4 = 16$, and so there is no way here of seeing his conjunction "and" in the role of the sign of identity. His functional decomposition of "$\vdash 2^4 = 16$" into "$x^4 = 16$" and, respectively, of "$\vdash 2^2 = 4$" into "$x^2 = 4$" appears to the left of "and" *not in place of* predication expressed in

(10.11) $$\vdash\!\!\!\begin{array}{l} \top\; x^4 = 16 \\ \llcorner\; x^2 = 4 \end{array}$$

in the form of the conditional, but only as component parts of this conditional, and hence also of the predication expressed by means of this conditional. But, as a corollary of the non-identity of the subject/predicate distinction, appearing in (10.11) as a decomposition into the antecedent and the consequent of the conditional, with the decomposition into function and argument in either of "$x^4 = 16$" or "$x^2 = 4$," the fact that (10.11) is true, which amounts to saying that the complex expression in (10.11) has instances, *turns up to be independent of whether its constituent functional expressions have instances themselves*. For even if they have such instances (as, for example, in Frege's substitution cases of "$2^2 = 4$" and "$2^4 = 16$"), the truth of (10.11) as a whole and the fact that its constituent functional expressions have instances (for some substituend of the value of x allows the respective concept to map it on to the True) happen to arise only coincidentally, with the truth of the former standing in no cause-effect relationship with the truth of the latter. In other words, their both being contingent facts does not at all imply that the first is contingent on the second. But then Hintikka's conjecture that "a first-order existential quantifier prefixed to an open formula says merely that the (usually complex) predicate defined by that open formula is not empty" is demonstrably false, for the truth of an open formula, asserted by means of a first-order existential quantifier being prefixed to it, arises quite independently of whether that open formula is empty or not. That is, that a complex predicate has instances or not turns up to be quite independent of whether we have some substituends for the gaps in that open formula or not. And, therefore, quite irrespectively not only of whether we pick an x such that it satisfies that formula or not, but regardless also of whether or not we have a winning strategy of how to do it. For whatever strategy of picking an x will then be simply irrelevant. As a corollary, no less irrelevant appears the whole framework of game-theoretical semantics (GTS), specifically in view of what Hintikka and Sandu conceptualise as a distinction between (i) games of verification and falsification (as *exemplified* by the semantical games of GTS), on the one hand, and (ii) games of formal proof (as *not* exemplified by the semantical games of GTS), on the other, cf.:

> The distinction between games of verification (i) and games of formal proof (ii) is important but is often overlooked. Yet the difference could not be more striking. Verification games serve to define truth; proof games aim at capturing all (and only) logical truths. Now truth and logical truth are categorially different. Truth presupposes a fixed interpretation and is relative to a single model. Logical truth means truth on any interpretation (in any model). Verification games are played among the objects one's language speaks of; proof games are played with pencil and paper (or on a computer).[65]

But, as we have seen from the foregoing analysis, the case of Frege's quantifier is far from being referential or objectual; rather, it makes a case of purely substitutional quantification, where, according to Quine, "we preserve distinctions between the true and false, as in truth-function logic itself, but we cease to depict the referential dimension."[66] Therefore, all it can do is, at best, quantifying into truth-functional connective positions, the latter being rendered, following Orenstein,[67] by something like $(\exists f)(pfp)$. So, the best one can hope for of Hintikka's game-theoretical rules to accomplish is to provide a semantical game just for $(\exists f)(pfp)$. It would be hard to envisage, however, how this game could be construed under the category of games of verification (i) rather than that of games of formal proof (ii).

---

[65] J. Hintikka, and G. Sandu, "Game-Theoretical Semantics," in Johan van Benthem, and Alice ter Meulen, eds., *Handbook of Logic and Language*, Elsevier, Amsterdam, 1997, p. 404.

[66] Alex Orenstein, "Towards a Philosophical Classification of Quantifiers," in A. Orenstein, and R. Stern, eds., *Developments in Semantics*, p. 91.

[67] *Ibid.*, p. 90.

CHAPTER TWO

THE PRINCIPLE OF IDENTITY AND ITS INSTANCES

> But by what right does such a transformation take place, in which concepts correspond to extensions of concepts, mutual subordination to equality? . . . We will have to assume an unprovable law here.
>
> *G. Frege, Posthumous Writings, p. 182.*

> Hence it is no exaggeration to say that to understand first-order logic is to understand the notion of a dependent quantifier. If my learned friend just quoted is right, if you do not understand the notion of a dependent quantifier you do not understand anything.
>
> *J. Hintikka, The Principles of Mathematics Revisited, p. 47.*

2.1. *The Aboutness of Propositions*

As has already been stated, for Frege "[w]ork in logic is, to a large extent, a struggle with the logical defects of language, and yet language remains for us an indispensable tool. *Only after logical work has been completed shall we possess a more perfect instrument.*[1] This was his reason for the construction of his *Begriffsschrift*, and the most conspicuous result of Frege's struggle with the logical imperfection of language manifested itself in his rejection, for the purposes of the construction of a more perfect logical instrument, of the subject/predicate distinction.

According to Frege, imperfection is being induced by the presence of man and all source of imperfection is being sought in the 'psychological' character of the distortion being imposed by man, which allegedly adulterates the genuine character of a sign *qua* sign and thus makes it imperfect. Signals of such 'psychological' kind of adulteration were relished in passive constructions whose difference from the corresponding factually synonymous active constructions is understood to be lying not in

[1] G. Frege, "My basic logical Insights," in *Posthumous Writings*, (H. Hermes, F. Kambartel, F. Kaulbach, eds., Basil Blackwell, Oxford, 1979), p. 252.

the sphere of purely 'material' meaning which appears to be the same but in the sphere of extralogical; hence, 'psychological' meaning.

In terms of Frege's most direct predecessor, Trendelenburg, it is this psychological increment which prevents ordinary language to fully realise the task of the linguistic sign and makes it imperfect and, thus, prevents the linguistic sign of being a genuine sign. According to Trendelenburg, ordinary language accomplishes the task of the linguistic sign only partially because, in ordinary language, 'only to a small extent is there an internal relation between the sign and the content of the signified idea'.[2] From this observation there arises the idea of a language which brings 'the shape of the sign in direct contact with the content of the concept'.[3] This kind of language is to be called a *Begriffsschrift*. It is partially realised in the symbolism of mathematics but even then the goal is not yet fully achieved; there is not yet a language in which the connection between signs and concepts is *logical*, and not merely psychological.

Thus Leibniz's and Trendelenburg's idea of a genuine linguistic sign as one which is completely free from all deterioration induced by the very presence of man appears to be a denial of Peirce's conception of a genuine sign where, if it is *genuine*, then it is genuine exactly *due to* this very presence. The real source of the conflict lies, however, not in psychological *vs.* non-psychological, but in the understanding of the nature of the bond between the sign as an object or a so-called sign-vehicle and the content which it signifies. But before going into this matter more closely, consider Frege's justification of his rejection of the distinction between subject and predicate:

> In order to justify this, let me observe that there are two ways in which the content of two judgments may differ; it may, or it may not, be the case that all inferences that can be drawn from the first judgment when combined with certain other ones can also be drawn from the second when combined with the same other judgments. The two propositions 'the Greeks defeated the Persians at Plataea' and 'the Persians were defeated by the Greeks at Plataea' differ in the former way; even if a slight difference of sense is discernible, the agreement in sense is preponderant. Now I call the part of the content that is the same in both the *conceptual content*.[4]

[2] A. Trendelenburg, "On Leibniz's Project of a Universal Characteristic", *Historische Beiträge zur Philosophie*, vol. 3, Berlin, 1867, p. 3.

[3] *Ibid.*

[4] *Begriffsschrift*, §3.

As Frege explains,

> When people say 'the subject is the concept with which the judgment is concerned,' this applies equally well to the object. Thus all that can be said is: 'the subject is the concept with which the judgment is chiefly concerned.' In language the place occupied by the subject in the word-order has the significance of a *specially important* place; it is where we put what we want the hearer to attend to specially. This may, e.g., have the purpose of indicating a relation between this judgment and others, and thus making it easier for the hearer to grasp the whole sequence of thought. All such aspects of language are merely results of the reciprocal action of speaker and hearer; e.g. the speaker takes account of what the hearer expects, and tries to set him upon the right track before actually uttering the judgment. In my formalized language there is nothing that corresponds; only that part of judgments which affects the *possible inferences* is taken into consideration.[5]

As we can see, the main antipsychologistic argument in Frege's exposition is based on purely logical grounds: 1) what belongs to the logical content is only that which affects possible inferences, 2) parsing into subject and predicate has nothing to do with possible inferences, therefore, 3) parsing into subject and predicate has nothing to do with logic. This, in Frege's terms, amounts to saying that what the subject/predicate distinction has to do with are purely psychological aspects of the meaning of the utterance. And the psychological aspects of the meaning of the sign are exactly those which distort the purely logical aspects of the meaning of the sign and make this sign logically defective. Hence, in a purely logical notation, they must be entirely dispensed with. The distinction between subject and predicate, being psychological, is not merely extralogical and thus logically excessive as serving purely communicative purposes which are not strictly logical: facilitating communication by regrouping subject and predicate using the active and passive forms of sentences is attained at the intolerable cost of adulterating the logical purity of what is being communicated.

What is being contaminated, by the use of the subject/predicate distinction, is in the first place the issue of what the statement, judgment, or utterance is about. In Frege's view, the real subject of an utterance performed with the use of natural language signs is not the meaning of any part of it, but the meaning of the utterance as a whole. As Frege explains,

> We may imagine a language in which the proposition 'Archimedes perished at the capture of Syracuse' would be expressed in the following way: 'the violent death of Archimedes at the capture of Syracuse is a fact.' You may

[5] *Ibid.*

> if you like distinguish subject and predicate even here; but the subject contains the whole content, and the only purpose of the predicate is to present this in the form of a judgment.[6]

In another place he adds:

> In the present context the only essential thing for us is that a different thought does not correspond to every difference in the words used, and that we have a means of deciding what is and what is not part of the thought, even though, with language constantly developing, it may at times be difficult to apply. The distinction between the active and the passive voice belongs here too. The sentences 'M gave document A to N', 'Document A was given by N to M', 'N received document A from M' express exactly the same thought; we learn not a whit more or less from any one of these sentences than we do from the others. Hence it is impossible that one of them should be true whilst another is false. It is the very same thing that is here capable of being true or false.... Although in actual speech it can certainly be very important where the attention is directed and where the stress falls, it is of no concern to logic.[7]

Here, thought or else conceptual content seems to be equated with the fact it expresses (cf. Frege's paraphrase of 'Archimedes perished at the capture of Syracuse' by 'the violent death of Archimedes at the capture of Syracuse is a fact'), and Frege's claim regarding logical irrelevance of stress and passive/active distinction turns up to be grounded on the factual equivalence holding between 'M gave document A to N' and its derivatives. In other words, their logical form remains the same because both the active sentence and its two passive counterparts, 'N was given document A by M' and 'Document A was given to N by M', as well as all its phonetic realizations amounting to

(1.11) It was M who gave document A to N
(1.12) It was N whom M gave document A
(1.13) It was document A that M gave to N

are about the same fact rendered by Frege's new subject/predicate parsing in something like 'that M gave document A to N is a fact'.

This proposal, however, runs into immediate difficulties in view of examples such as (1.14):

(1.14) I told three of the stories to many of the men.

---

[6] *Ibid.*

[7] G. Frege, "Logic", in *Posthumous Writings*, p. 141.

Sentence (1.14), discussed at length in Jackendoff (1972), Hintikka (1974), and Fauconnier (1975), provides striking counterevidence to Frege's claim, since depending on where the stress falls Frege's sentential predicate of the form '... is a fact' will now be seen as making assertions not about one and the same fact but about *different* facts. A closer examination of (1.14) shows perfectly well that Frege's conjecture—that different stress placement only shifts attention to different parts of what is essentially one and the same logical content—turns up to be demonstrably false. The shift of stress in effect does not produce merely varying psychological accents on what is essentially the same objective content as in Frege's example 'M gave document A to N' but, on the contrary, serves to differentiate between what appears to be essentially different objects depending on whether (a) there are many of the men who are each told three stories (the stories may be different for each man); (b) there are three particular stories such that each is told to many men (the men may be different for each story); or (c) there are three particular stories and a particular group of many men such that the stories are told to the men.[8] A little bit closer consideration will make it clear that Frege's example, 'M gave document A to N', which he uses for the justification of the unity of a proposition regardless of the placement of stress, turns up to be only occasionally synonymous with (1.11)—(1.13), but—generally—*ambiguous* between the structural meanings rendered by (1.11)—(1.13). The synonymity of 'M gave document A to N' when pronounced with different stress presents only a particular case in the fulfilment of much more general conditions, and instantiating no more but a limiting case of these conditions does not, by itself, determine any general rule. In Frege's examples they happen to be synonymous not—as it is conjectured—due to a profoundly logical fact: the fact that they are synonymous happens to be trivially contingent, depending on the absence of quantifier expressions in the structures intended as an unshakable proof of a profound logical truth. Frege's justification of his idea of logical irrelevance of sentence stress thus makes a clear example of a *non sequitur*. By way of logical consequence, as a *non sequitur* presents itself also the representation (and the very idea of such a representation) of the meaning differences between various readings of (1.14) in terms of different quantifier scopes due to different configurations of quantifiers, prefixed to an invariant structural representation of

[8] Cf. R. S. Jackendoff, *Semantic Interpretation in Generative Grammar*; J. Hintikka, "Quantifiers vs. Quantification Theory", *Linguistic Inquiry* 5, 1974, 153–77; G. Fauconnier, "Do Quantifiers Branch?", *Linguistic Inquiry* 6, 1975, 555–78.

one and the same form, $R(x, y, z)$. As the result, discerning the meaning variations of $R(x, y, z)$ in terms of scope distinctions leaves at least one of the interpretations of (1.14) entirely unaccountable.

Furthermore, not only stress variations, but active/passive distinctions, too, appear in fact to have nothing to do with the allegedly strictly psychological aspects of communication that should be of no concern to logic. Attention to counterevidence in this respect was first drawn by Chomsky who in *Syntactic Structures* (1957) made an observation to the effect that

> In contradiction to (117vi) [an active sentence and the corresponding passive are synonymous], we can describe circumstances in which a 'quantificational' sentence such as "everyone in the room knows at least two languages" may be true, while the corresponding passive "at least two languages are known by everyone in the room" is false, under the normal interpretation of these sentences—e.g., if one person in the room knows only French and German, and another only Spanish and Italian. This indicates that not even the weakest semantic relation (factual equivalence) holds in general between active and passive.[9]

But, since 'quantificational' sentences appear to be indicating that active and passive are not synonymous, the question that naturally arises is whether we should treat such 'quantificational' sentences as exceptions to the rule or, on the contrary, whether it is the rule itself that is only applicable to exceptions (like those exemplified in Frege's elucidations of his language of 'pure thought'). The then wide acceptance of Chomsky's views on the theory of syntax notwithstanding, the former of these options turns up as strikingly counterintuitive and in addition in effect anti-Fregean.

Assessing his assertions of (117), and among them also his assertion (117vi) [an active sentence and the corresponding passive are synonymous] in terms of lying somewhere within the range between being 'not wholly false' and being 'very nearly true'(?!), Chomsky however attributes the source of what he considers to be merely a nuisance, a sort of quantificational anomaly, and an indication of semantic 'misbehaviour' of quantifiers, to the fact that "only imperfect correspondences hold between formal and semantic features in language,"[10] thus relegating quantificational sentences to a dismally marginal place of what he holds to be only semantic inadequacies in the relations between sentences. This point of view[11]

---

[9] Noam Chomsky, *Syntactic Structures*, Mouton, 1957, pp. 100–101.

[10] *Ibid.*, p. 101.

[11] ... which later gave rise to the so-called 'problem of quantifiers' emerging in the context of an already historical discussion of 'whether transformations can change meaning', cf., e.g., Hintikka (1974).

cannot nevertheless be taken as anything but manifestly counterintuitive because it is exactly quantificational sentences that are long known to be making the central core of all logical expositions whenever the issue of inferential relations between sentences is involved, whereas statements involving proper names like 'Socrates is a man' on the contrary are generally known to be playing excessively marginal role—appearing, if at all, in expositions of the syllogistic mood known as "Barbara" only due to their implicit identification with *universal* statements;[12] hence, owing to the tacit treatment of them as *quantificational* sentences. So, singular statements like 'Caesar Gallos vicit Galli' and 'Galli a Caesare victi sunt' which, following the exposition in Thrane,[13] are both revealing the same logical content, as in *That Caesar defeated the Gauls is a fact*, in Aristotle's view would have indeed very little, if anything, to do with syllogistic inference—which was exactly the type of inference that Frege was explicitly referring to when he wrote about the alleged logical identity of his two judgments, 'the Greeks defeated the Persians at Plataea' and 'the Persians were defeated by the Greeks at Plataea', with respect to the inferences that may be drawn from these judgments when combined with certain other ones. Taking this into account, it does not seem convincing at all that it is exactly Frege's examples involving proper names as in 'M gave document A to N'—rather than 'quantificational' statements such as (1.14)—should then be regarded as providing *more basic* facts, especially when what is at issue is specifically relevance with respect to possible *inferences*.

Furthermore, the first option may be shown as inconsistent not only in the light of Frege's dictum according to which what belongs to the content is exclusively what is relevant to the validity of inferences, but also against the background of Frege's insistence on that his analysis into function and argument, being intrinsically arbitrary, comes to acquire logical (*inhaltlich*) significance exactly when sentences that are involved happen to be of the kind that we would now call 'quantificational'. Because for Frege,

> ... the different ways in which the same conceptual content can be considered as a function of this or that argument have no importance so long as function and argument are completely determinate. But if the argument becomes *indeterminate*, as in the judgment: "Whatever arbitrary positive integer we take as argument for 'being representable as the sum of four squares', the

[12] Cf. M. V. Aldridge, *English Quantifiers. A Study of Quantifying Expressions in Linguistic Science and Modern English Usage*, p. 3; D. Mitchell, *An Introduction to Logic*, p. 43.

[13] T. Thrane, *Referential-Semantic Analysis: Aspects of a Theory of Linguistic Reference*, p. 36.

> [resulting] proposition is always true.", then the distinction between function and argument acquires a *substantive* {*inhaltlich*} significance.[14]

But from this it would naturally follow, as a consequence, that it is exactly in a 'quantificational' sentence, but never in a sentence containing only logical proper names,

> ... the whole splits up into *function* and *argument* according to its own content, and not just according to our way of looking at it.[15]

In effect, what Frege claims by this is that, for logical purposes, the way that we parse 'M gave document A to N' into function and its arguments by way of representing it as something like '$R(x, y, z)$' is not important at all—and thus, in effect, logically irrelevant. Consequently, what the synonymity of all structural and prosodic versions of 'M gave document A to N' does in fact confirm is only the validity of Frege's own principle that the whole splits up into *function* and *argument* according to our way of looking at the content (in this case, owing to the presence of proper names M, A, and N, due to which the required indefiniteness is lacking). But from this it does not follow in the least that the content we are looking at is in all instances the same. For, if the referents of M, A, and N were indeed deemed indefinite, the whole would then split into function and argument according to its *quantificational* content like the one in example (1.14). But no matter if we consider Jackendoff's or Hintikka's quantificational examples like (1.14), or examples like Chomsky's in (1.15):

(1.15) Everyone in the room knows at least two languages
(1.16) At least two languages are known by everyone in the room

or else their logically more adaptable counterparts in (1.17–1.18)

(1.17) Everyone loves someone
(1.18) Someone is loved by everyone

it is immediately evident that what in deed *is* logically relevant turns up to be the parsing into function and arguments exactly in those quantificational examples that provide counterevidence to Frege's claim as to the logical irrelevance of active/passive distinctions and—what follows

[14] *Begriffsschrift*, §9.
[15] *Ibid.*

as a mere corollary in view of Frege's examples 'the Greeks defeated the Persians at Plataea' and 'the Persians were defeated by the Greeks at Plataea'—they also provide counterevidence to Frege's claim as to the logical irrelevance of the subject/predicate distinction as well. What this basically means is that Frege's claim that the distinction between function and argument acquires a *substantive* {*inhaltlich*} significance if and only if the argument becomes *indeterminate*[16] is, for all ends and purposes, incompatible and irreconcilable with what he claims elsewhere—that the distinction between subject and predicate is logically irrelevant.

2.2. *Frege, Euler, and Schröder's* Quaternio Terminorum

In the light of what has been said above, it will be seen that what is indeed of prime importance lies not in Frege's demonstrations of how single concepts (in effect, general names[17]) may be analysed into function and argument(s) and, as a consequence, not in the explication, abounding in the literature, of the atomic sentences with complex (both linear as well as branching) quantifier prefixes of the form $QxQn \ldots Qk(Fx, n, \ldots k)$ whose logical relevance will only be shown as fictitious, but in the consideration of how Frege's functional analysis of *quantificational* sentences fares in the face of his reformulation—in the form of the conditional—of the traditional subject/predicate distinction in categorical statements.

Conspicuous in this respect is Frege's discussion of the Euler diagrams, which he describes in his review of Schröder's *Vorlesungen über die Algebra der Logik* as "a lame analogy for logical relations." Criticising "the superficial view as to the concept (one might call it a mechanical or quantitative view) that comes out also in Euler's diagrams," according to which 'some numbers' in

(2.21) Some numbers are prime

[16] ... which is not the case with 'M gave document A to N', because in 'M gave document A to N' the values of M, A, and N are unknown but *determinate* values, and *not* such values that we know them to be *indeterminate*, as in "Whatever arbitrary positive integer we take as argument for 'being representable as the sum of four squares', the [resulting] proposition is always true." The use of letters in 'M gave document A to N' does not express generality in much the same way as generality is not expressed, according to Frege's own testimony (see p. 21 above), by the use of the letter x in '$x^2 - 4x + 3 = 0$'.

[17] ... which, as Quine (1972:218) as well as Mill (1843:18) before him explain, may be said to be *true* with respect to those objects, of which they are names, otherwise false.

is regarded as the subject, and so is 'all bodies', or the concept *bodies* in its full extension, in

(2.22) All bodies are heavy

Frege argues that so far as the sense of the words 'some' or 'all' goes "these words must be counted in with the predicative part of the sentence."[18]

So far, these considerations seem to be in accord with the conception of the quantifier as a second-order function taking a first-order function as its argument which—as a concept of level-*n*—is also said to be *falling within* the quantifier as a concept of level-(*n*+1). And since this relation of *falling within*, which Bell calls Relation III,[19] seems to be analogous to Relation I, that of an object *falling under* a concept, this invites a hasty conclusion that what the quantifier as the concept of level-(*n*+1) asserts about the concept of level-*n* is that the latter has instances and thus says something about the objects falling under that concept. This conclusion is, however, unwarranted according to Frege's own standards, who in his review of Husserl's *Philosophie der Arithmetik* writes that

> The so-called common name—which would be better named 'concept word'—has nothing to do with objects directly, but stands for a concept. Under this concept objects may fall; but it may also be empty, and this does not stop the concept word from standing for something. I have already given an adequate exposition of this point in §47 of my *Foundations of Arithmetic*. It is surely clear that when anyone uses the sentence 'all men are mortal' he does not want to assert something about some Chief Akpanya, of whom perhaps he has never heard.[20]

But if so, then what does anyone want to assert about if not about any of the objects falling under the concept *man* and thus not about Relation I? Indubitably, this turns out to be Relation II, thus—not a *sub*-relation that holds between a concept and an object falling under this concept but, instead, quite a different kind of relation, namely one that holds between two concepts of the same level, which Frege describes as a *subter*-relation. For, as Frege makes it quite explicit with regard to his examples in (2.21–2.22):

[18] G. Frege, Kritische Beleuchtung einiger Punkte in E. Schröders Vorlesungen über die Algebra der Logik. In *Archiv für systematische Philosophie*, 1, p. 441n.

[19] D. Bell, *Frege's Theory of Judgement*, p. 35.

[20] G. Frege, Review of E. Husserl, Philosophie der Arithmetik. In *Zeitschrift für Philosophie und philosophische Kritik*, vol. 103, pp. 326–27.

> The word 'some' states a relation that holds (in our example) between the concepts 'number' and 'prime.' Similarly 'all' in the second example states a relation between the concepts *bodies* and *heavy*. An expression answering better to the logical structure is: 'bodies are universally heavy.'[21]

But, since 'some' and 'all' both state a relation that holds between *concepts*, the necessary condition for this relation to take place is that the concept within which this relation occurs must be, of necessity, *complex*. What Frege says about this is, in effect, that quantification involves a relation of subordination between concepts, but never a relation of an object falling under a concept. He clearly points out that *some* does not belong to *man* taken as a single concept, and what it has to do with is not the *sub* relation of an object falling under a concept, but the *subter* relation of a concept being subordinate to another concept of the same level. Surely, a concept is an agent in both kinds of relationship—that of concept-concept as well as that of concept-object, but what the *some* in *some numbers are primes* is connected with is not Relation I within a simple concept, but Relation II within a concept which, of sheer necessity, must be inevitably *complex*. Because, according to Frege, there can be no quantificational relation within *some men*, since it would necessarily involve a relationship between the two concepts, *some men* and *men*, but there exists no such concept as *some men* that could be regarded as being subordinate to the concept *men*, cf.:

> Herr Schröder gives an example (p. 180) of *quaternio terminorum* that arises because the expression 'some gentlemen' does not always designate the same part of the class of gentlemen. Accordingly such an expression would have to be rejected as ambiguous; and it must, in fact, be rejected if one regards it (like the author, p. 150) as designating a class that consists of 'some' gentlemen. Of course what I am here rejecting is not the particular judgment, but only a wrong conception of it.[22]

The wrong conception here is that of *some* forming a concept together with *gentlemen*. The subject, in Frege's terms, is not *some gentlemen*, but the whole judgeable content which, in order to be judgeable, must contain also another concept along with that of *gentlemen*, with *some* appearing as a predicate asserting something about this judgeable content as a whole. Thus, the content to which the quantifier applies must already *be*

---

[21] G. Frege, Kritische Beleuchtung einiger Punkte in E. Schröders Vorlesungen über die Algebra der Logik. In *Archiv für systematische Philosophie*, 1, p. 441n.

[22] *Ibid.*

judgeable, i.e. it must be judgeable *prior* to the application of the quantifier; hence, it cannot consist of a single concept no matter be it saturated or not. (In the first case, the quantifier is inapplicable on general grounds, and the second case is exactly the case when there is no judgeable content to which a quantifier could be applied as a higher-order predicator.) But from this it, however, follows that rendering the logical form of *some gentlemen* as something like $\exists xGx$ we would commit the same mistake that Frege addressed in his criticism of Schröder—that of making no distinction between the logically relevantly different *sub-* and *subter*-relations. Besides, in view "of *quaternio terminorum* that arises because the expression 'some gentlemen' does not always designate the same part of the class of gentlemen," our representation in $\exists xGx$ (or, generally, in $QxFx$) which we would then treat as a representation of 'a second-level concept within which a first-level concept falls' will be a representation of something that in Frege's terms is not a concept at all. Of course, the resulting paradox that a second-level concept is not a concept can be quickly resolved by stating that a second-order concept is not a concept, but a proposition. But *some gentlemen* is not a proposition, either.

Thus, it becomes pretty evident that all what the inscription in $G(x)$ requires in order to acquire meaning boils down in principle to obtaining, by '$G(x)$', the status of an element of a truth-functional context, whose necessity arises directly from Frege's principle, "never to ask for the meaning of a word in isolation, but only in the context of a proposition."[23] But the required context appears to be wholly independent of the context of quantification, which is already seen in Frege's own explication of quantificational logical structure of (2.22) in 'bodies are universally heavy' where the proposition that taken as a whole can be said to have a sense is 'bodies are heavy', and it surely is not the proposition 'bodies are universally', but the one that—according to Frege—confers on its parts also their content. This means that an imputation, by means of Frege's quantifier, of truth-functional context on what Frege depicts as an unsaturated concept is only fictional, for the quantifier turns up to be not truth-generating, but itself parasitic on a truth-functional context arising only when the number of concepts in question is no less than two, quite regardless of whether their link is considered as a connection between subject and predicate, and thus coined in terms of traditional logic, or else as a connection, by means of the conditional as Frege's expression of *logical* generality, between the

[23] G. Frege, *Die Grundlagen der Arithmetik*, p. 71.

two so-called open sentences which by themselves are no more truth-functional than the concepts they are deemed to represent.

2.3. *Ockham and Truth in Equation*

In so far as there is a purpose, which Frege definitely has, in redefining the issue of the aboutness of general propositions, Frege's agenda amounts to saying that insofar as the proposition is general, then it is not represented by a simple concept—not even if the quantifier as a second-order concept takes it as its argument—for the first-order concept in question should then necessarily appear in the form of the conditional in which it is important that both its parts should contain equiform letters (variables) so that they would "cross-refer to one another."[24] This cross-reference, for Frege, is a sufficient and necessary condition that both parts of the conditional must fulfil in order for the conditional to be (universally) true. Thus, the aboutness of the quantifier as the second-order predicate does not amount to stating that the lower-order predicate (concept) has instances and thus there are (or there can be found, as a result of the application of some of Hintikka's game-theoretic strategies) objects such that they are subsumed by the (usually complex) concept, for what Frege's expression of generality appearing in front of 'If *a* is a man, *a* is mortal' in

(3.23) ├─𝔞─ If *a* is a man, *a* is mortal

and, accordingly, in

(3.231) ├─𝔞─┬─ *a* is mortal
└─ *a* is a man

in effect states is nothing more and nothing less than that the *a*'s in both the antecedent and the consequent of the conditional are co-referential.

In doing so, Frege is apparently justified by what Ockham wrote on truth—since in the first place it was truth, not (material) reference, that was of prime concern to Frege—in his remarkable passage:

> ... for the truth of the proposition 'This is an angel' it is not required that this common term 'angel' be really the same as that which has the position

[24] G. Frege, "Logical Generality", in *Posthumous Writings*, p. 260.

> of subject in this proposition . . .; but it is sufficient and necessary that subject and predicate should stand for [*supponant*] the same thing.[25]

Frege takes due notice of this necessary and sufficient requirement for the truth of the proposition, which is seen as the very first point that he is making in *Begriffsschrift* with regard to his use of variables, i.e. that the (variable) letter must retain in a given context the same meaning once given to it. But this requirement in no case applies to the sameness of meaning between the Gothic letter placed in the concavity and the same letter appearing in the first-order functional expression. For, according to Frege, whenever we give whatever meaning to a variable, then we are dealing with the first-order functional expression alone because the concavity with the Gothic letter in it (thus, quantifier prefix as such) must then disappear. Thus, the idea that is basic to Frege's logic is not Frege's idea of quantifier dependence, as Hintikka thinks and what makes him seek the true basis of quantificational logic in a wider context where Frege's idea appears as only a limiting case of informational independence, for the object of Frege's concern is directly the opposite to that of Hintikka's. Frege's basic concern is *not with the mutual interplay of various variables* when what is given is the same concept with its meaning taken as invariant but, on the contrary, his basic concern is with *the mutual interplay of various concepts* when the main requirement is that the meaning, once given to a variable, remains invariant throughout the whole context of this mutual interplay. For the main requirement for obtaining truth is not that the two terms occurring in the context of a proposition (and thus the concepts that they represent) should be identical, but that the *thing* for which these terms stand should necessarily be the same and thus, in Frege's terms, what must be the same is the object that falls under the concepts represented by the terms in question.

This Frege finds to be exactly the case in algebraic equations, when what is being equated happens to be various numerical terms representing functions, where the necessary precondition for the truth of the equation also amounts to the identity of numerical values and thus—in a sense—objects being substituted in place of variables. Nevertheless, being just a necessary requirement with respect to algebraic equations, this however does not happen to be also a *sufficient* requirement for obtaining the truth of the proposition expressed by an algebraic equation, for what the iden-

[25] Ockham, *Summa Totius Logicae*, II, c. ii. Cf. also St. Thomas, *Summa Theologica*, I, q. 13 a. 12.

tity of numerical values under all substitutions can only yield is *a* truth value, but not necessarily the truth value True, as in Frege's example in

$$(3.24)\quad (x + 1)^2 = 2(x + 1)$$

in which Frege's logical function always has "the same value for the same argument, *viz.* the True for the arguments −1 and +1, and the False for all other arguments."[26] Besides, the truth of these equations seems to be evidently contingent since the necessary requirement appears to be that these values should only be numerical—otherwise the equation is false or else nonsensical. Moreover, the functional expressions to the left and to the right side of the sign of equality in the numerical equations do not appear as concepts, for what receives the truth value True or False turns up to be the whole of the equation but neither the part to the left nor to the right of the equals-sign. Thus, the application of Ockam's idea directly to the mathematical expressions in the form of an equation ends up in an immediate failure, which gives Frege reasons to think that what is at fault, and what thus happens to be essentially the cause of this failure, is the interpretation of the equals-sign as a logical copula, cf.:

> If a function is completed by a number so as to yield a number, the second is called the value for the first as argument. People have got used to reading the equation '$y = f(x)$' as 'y is a function of x.' There are two mistakes here: first, rendering the *equals*-sign as a copula; secondly, confusing the function with its value for an argument.[27]

and, as a corollary, what is at fault happens to be the subject/predicate distinction itself, of which the copula is intrinsically a part. To put Ockham's idea to work, Frege has no way out but to put what he views as a subordination of concepts (equivalent to Ockham's relationship between two common terms within a proposition) into the context of a relationship which, of necessity, could not be regarded as a heterogeneous relation between a function, $f(x)$, and its value, y, and even less as an equally heterogeneous relation like the one between an argument, x, and the function of that argument in $f(\ )$ that we see in Frege's example above. Thus the relation in question should necessarily be homogeneous, and not heterogeneous like the one that can be found within a *simple* concept as such and, at the same time—as it follows from Frege's remark—this relation cannot

[26] G. Frege, *Funktion und Begriff*, p. 16.
[27] G. Frege, *Was ist eine Funktion*, p. 665.

be rendered by means of the copula. To find the appropriate context, Frege uses the distinction in his description of concepts that he poses between the notions of 'property' on the one hand, and 'mark', or 'characteristic', on the other. According to Frege, both are words that "signify relations, in sentences like 'Φ is a property of Γ' and 'Φ is a mark of Ω'."[28]

In the first of these relations, the 'Φ' that symbolises a concept appears in a heterogeneous relation with 'Γ', the latter being the symbol standing for an object that falls under that concept, whereas the Φ in the second relation appears to be standing in a homogeneous relation with another concept of which, as a partial concept, it is a characteristic mark. As Frege explains in his letter to Liebman:

> Concepts are usually composed of partial concepts, the characteristic marks. *Black silk cloth* has the characteristic marks *black*, *silk* and *cloth*. An object which falls under this concept has these characteristic marks as its properties."[29]

Thus, an object falling under the characteristic mark of the concept in 'Φ is a property of Γ' will then necessarily be falling under the more complex concept Ω, of which the characteristic mark or partial concept Φ is only a constituent part in much the same way as *black* appears graphically as a constituent part of the inscription in *black silk cloth*. In other words, the truth of, say, *This black silk cloth is black* obtains due to the fact that an object falling under the superordinate concept Φ which is represented as *black* to the right of *is* also falls under the subordinate concept Ω which is *black silk cloth* to the left of *is* just because of the fact that Ω, which is *black silk cloth*, also contains Φ, i.e. *black*, as one of its characteristic marks. Which shows that what appears as a copula in *This black silk cloth is black* may now be conceived not as a copula standing in a relation between subject and predicate, but as an equals-sign indicating a reversible equation-like relation between the characteristic mark, Φ, occurring to the right and to the left of *is* in *This black silk cloth is black* in the manner of '*This Φ silk cloth is Φ*' where the essence of the analytic truth in *This black silk cloth is black* is expressed by means of the evident analyticity of $\Phi = \Phi$. As a consequence, since whatever is a mark of a concept is a property of an object which falls under that concept, dealing with the same characteristic mark to the right and to the left of *is* means that we are dealing with something which is standing for the same thing, but which

---

[28] G. Frege, *Über Begriff und Gegenstand*, p. 201.

[29] Letter to Liebman, in G. Frege, *Philosophical and Mathematical Correspondence*, p. 92.

occurs not in the context of the subject/predicate distinction to which Ockham in his passage on truth refers, but in the context of the familiar mathematical generality of an algebraic equation in which, as Frege thinks, we deal with logical generality of which an algebraic equation is just an instance. By analogy, in the case of the subordination of a first-level concept under a first-order concept in examples such as 'All horses are four-legged animals',[30] or 'All mammals have red blood',[31] or 'All mammals are land-dwellers',[32] or 'All squares are rectangles',[33] truth—according to Frege—obtains owing to the fact that "The *characteristic marks* of the superordinate concept (rectangle) are also *characteristic marks* of the subordinate one (square)."[34] And this is expressly what Frege means when he writes, in *On Concept and Object*, that

> When I wrote my *Grundlagen der Arithmetik*, I had not yet made the distinction between sense and reference; and so, under the expression 'a possible content of judgment,' I was combining what I now designate by the distinctive words 'thought' and 'truth-value.' Consequently I no longer entirely approve of the explanation I then gave, as regards its wording; my view is, however, still essentially the same.[35]

Moreover, it is exactly due to this implicit specific identity of sense and reference with regard to which Frege's view remains *essentially* the same that, even though linguistically "... the words 'all,' 'any,' 'no,' 'some,' are prefixed to concept words," he goes on expressly to conclude that

> In universal and particular affirmative and negative sentences, we are expressing relations between concepts; we use these words to indicate a special kind of relation. They are thus, logically speaking, not to be more closely associated with the concept-words that follow them, but are to be related to the sentence as a whole.[36]

Clearly, this 'sentence as a whole' cannot be seen as a single concept construed entirely and solely in primitive terms of 'object-concept' relationship, to which Frege's exposition of *Sinn* and *Bedeutung* solely applies. Not even if the number of objects falling under this concept turns up to be more than one, as in the case of polyadic predication to which the theory

[30] G. Frege, *Über Begriff und Gegenstand*, p. 196.
[31] *Ibid.*, p. 197.
[32] *Ibid.*, p. 198.
[33] Letter to Liebman, in G. Frege, *Philosophical and Mathematical Correspondence*, p. 93.
[34] *Ibid.*
[35] G. Frege, *Über Begriff und Gegenstand*, p. 198.
[36] *Ibid.*

of branching quantifier prefixation applies. The logical essence of 'the sentence as a whole'—in the way Frege sees it—is in the subordination of concepts that this sentence expresses. And what makes this sentence and thus the subordination of concepts that it expresses true—the equation in $\Phi = \Phi$—makes also the exact sense in which Frege's distinction between *Sinn* and *Bedeutung* does not matter. Because the distinction between Frege's *Sinn* and *Bedeutung* turns up to have nothing to do with the relevant equation of the form $\Phi = \Phi$ that Frege finds to be involved in the relationship between the subordinate and the superordinate concepts as a necessary as well as sufficient condition for this relationship to be assessed as a whole as rendering the truth-value True.

2.4. *Frege's Improvement on Kant: Synthetic Statements as Kind of Analytic*

The implicit containment of $\Phi = \Phi$ in the relationship of concept subordination is, for Frege, the issue of the containment of $a = a$ in $a = b$. The first of which—being exactly the form of $\Phi = \Phi$—is analytic and thus insensitive to the distinction between *Sinn* and *Bedeutung*, whereas the second—representing the general form of concept subordination as such—is synthetic, to which Frege's distinction fully applies. For Frege this issue forms the backbone of his philosophy, in which he improves upon Kant's distinction between synthetic and analytic. For what Frege sets as his ultimate task, and which also makes the philosophical motto of his programme of logicism, is that all synthetic judgments of arithmetic are in fact analytic, and thus their cognitive value is connected with knowledge that is not empirical but purely *a priori*.

In Kant's definition, a judgement of the subject-predicate form is analytic if there is nothing contained in the predicate, which was not already thought in the concept of the subject; otherwise it is synthetic. The idea of this distinction seems to be derived from Locke's discussion of trifling and instructive propositions. Contrary to Kant, Frege does not view this distinction as clearly cut as in Kant's case. Frege's concern is to show that arithmetical statements that have the form of synthetic judgements, i.e. judgements in which the predicate (Frege's superordinate concept) *does* contain something which was not already thought in the concept of the subject (Frege's subordinate concept), are not stemming from empirical observation, as synthetic judgements are generally supposed to be, but can be traced to certain logical truths which are not synthetic but analytic. A clear case of Kant's synthetic judgement we find in, e.g., *all men are mortal*, which is evidently synthetic because, as it immediately follows from Euler's diagrams, what is thought in the concept *mortal* need not necessarily be *men* only. But it will not be so clear if we, following Frege, insist that *all men* is only a psychological subject which is such because we want to draw the attention of the hearer to some part of what our thought is about, whereas the full thought-content of *all men are mortal* involves not only that but also what is thought in the concept *mortal*, and thus involves objects such as tigers, birds, butterflies, and so forth. And, since their distinctive character notwithstanding the words 'thought' and 'truth-value' both mean the possible content of judgment, and in this respect what they mean is in Frege's view *essentially* the same; we can show that what we think in *all men are mortal* are not *men* alone, but also everything that is *mortal*. All we then need is to render the old subject/predicate distinction by means of the conditional in which *men* and *mortal* will appear in the form of logical functions taking the True and the False as their values. Since our primitive functions turn up to be true with respect to all objects that fall under the concept for which the function stands and whose properties the concept represents, then the thesis that what we think is the extension of *mortal* and not only of *men* will logically follow from the truth of the conditional (which is the truth of the whole of our though-content) arising in the case of the consequent (= *mortal*) being true and the antecedent (= *men*), false.

The next step that we need to accomplish in order to show the cognate character of the content of our synthetic thinking with the logical content of purely analytic thought will then be the recognition, along with Frege, of every declarative sentence as a proper name and not only the True but alternately also the False as its reference, cf.:

> Every declarative sentence concerned with the reference of its words is therefore to be regarded as a proper name, and its reference, if it has one, is either the True or the False. These two objects are recognized, if only implicitly, by everybody who judges something to be true—and so even by a sceptic.[37]

Thus, taken as "granted that every grammatically well-formed expression representing a proper name always has a sense,"[38] we will now have the thought as the sense of every declarative sentence expressed, by contrast, not in the subject-predicate form but in the form of the conditional, and

[37] G. Frege, *Über Sinn und Bedeutung*, p. 34.
[38] *Ibid.*, p. 28.

the reference of the conditional in question will be the True even if what we think in a sentence as a whole happens to be objects falling neither under the concept *men*, nor under the concept *mortal*. Having this in mind, the conditional is then conceivable as a means capable of providing the proof that the synthetic judgments of arithmetic are in fact analytic, and the conditional will therefore have to be put in place of the coinage of the subordination of concepts in terms of subject and predicate, since the attainment by the conditional of the same truth-value True regardless of whether the concepts *men* and *mortal* appear as either both true or both false with respect to relevant objects that may be either falling or else not falling under them indicates that in both cases we are dealing with the same reference of the sentence as a whole, and thus with the same thought expressed by that sentence. This follows directly from Frege's train of thought in his programmatic paper *On Sense and Reference*, where he concludes that

> If now the truth value of a sentence is its reference, then on the one hand all true sentences have the same reference and so, on the other hand, do all false sentences. From this we see that in the reference of the sentence all that is specific is obliterated.[39]

It will now be seen that placing the reference of the sentence in terms of its truth value as a starting point of Frege's analysis of the structure of propositions, which seemingly serves the ultimate purpose of attaining the expression of generality, has a more specific purpose in view—*namely that of obliterating all that is specific*. And, in the first place, obliterating all that the sentence is specifically about as reflected by the subject-predicate parsing. For only through a denial of the status of the equals-sign as a copula can one forcibly claim that the logical content of an algebraic equation is completely reversible and, therefore, that it is nothing but strictly analytic. Or else derivable, via truth-preserving transformations, from statements that are strictly analytic. For analytic knowledge is free from any need for empirical observation; hence, not requiring any power of observation at all.

It does not take much effort to observe, however, the strict coincidence between what Frege wants to prove in carrying out his logicistic programme and what, at the very start, he assumes as an indispensable instrument for providing such proof. Frege most apparently does not see

[39] *Ibid.*, p. 35.

any vicious circle here, because for him the necessity of rendering the basic logical structure of a statement in terms of the conditional rather than in terms of subject-predicate parsing does not follow from the ultimate necessities of his logicism: for him it seems to follow from the need of overcoming what he sees as a psychological burden that in his view renders language as only a distorted representation of thought and its grammar as an inadequate means of representing the real structure of thought. Nevertheless, regardless of the extent of antipsychologistic rhetoric that Frege employs in his argumentation, it can already be clearly seen that Frege's real motives in introducing the conditional are pro-logicistic rather than antipsychologistic. For the reason that Frege believes that the subject-predicate parsing is logically irrelevant stems not so much from his antipsychologistic convictions, as from the fact that it lacks the required reversibility that he wants since he finds it as a characteristic feature of algebraic equations. And even though subject and predicate may be thought as reversible, as it is actually being done in Aristotelian logic, they are still not identities in the sense that Frege requires so that his goal could be attained. And this is exactly to this end that he thinks is not to be achieved otherwise that he attends the issue of what he assumes as basic synonymity of active and passive. But, quite on the contrary to what Frege is constantly saying on the topic of antipsychologism, his argument on the issue of concept subordination is levelled not against the basic asymmetry arising as he thinks from the psychologistic treatment of concepts in terms of subject/predicate distinction, but against the equally basic asymmetry of his own object/concept distinction and thus the relation of "falling under" with which concept subordination, as he asserts on a number of occasions, should not be confused.

Sluga seems to be absolutely right in suggesting, that Frege's concern in his essay *On Sense and Reference* with the nature of *synthetic* or potentially knowledge-extending identities specifying ordinary objects should be understood as a stage in the working out of his mature account of analytic (logically true) identities required for the adequate specification of the logical objects treated in the *Grundgesetze*.[40] For what Frege is talking about is only *prima facie* a kind of close affinity that he poses to exist between physical objects, numbers and truth-values, which are all identical in that respect in which they appear to be referents of linguistic expressions. This is exactly the reason why he explicitly affirms

[40] Cf. Robert B. Brandom, 'Frege's Technical Concepts', in *Frege Synthesized*, p. 280.

on a number of occasions that the two expressions, '2^2' and '2 + 2' express different senses. But this talk serves only expository purposes because at bottom for Frege ordinary objects and numbers are not quite alike. For the purposes of the *Grundgesetze* numbers must be treated in a special way, and their identities in fact *are* logical identities, appearing in the way they are given in the very definition of the quantifier in *Function and Concept*, where the meaning of the quantifier in $—\overset{\mathfrak{a}}{\smile}— f(a)$ is given not in the way it was originally introduced in *Begriffsschrift*, but as the meaning of an identity that is to the same extent numerical as at the same time also logical, in $\vdash\overset{\mathfrak{a}}{\smile}— a = a$.[41] However, the introduction, along with the True, also of the truth-value False in the role of referents of linguistic expressions, makes physical objects as such quite dispensable, for this puts us in a position to speak about them not only without necessarily naming them (Quine's favourite turn of phrase), but—because of the properties of the conditional—also quite regardless of whether they in fact fall under respective concepts or not, and in effect regardless of their very existence and, thus, not to speak about them at all. In effect, all that Frege achieves by this ingenious step is that he brings to one's view not the objects that language is about but utterly abstract logical structures, a sort of logical furniture, to which these objects do not—and cannot—properly belong. To pursue Goldfarb's metaphor a bit further, this logical furniture turns up to be effectively locked from whatever objects with respect to which it was intended to serve as a placeholder. And in very much the same way that as Frege noticed we cannot speak about the referents of clauses for all we can speak about as referents are at utmost clauses themselves, we also cannot speak about objects, for the objects that we are only able to speak about are not objects that are falling under the first-level concept but only objects formed exclusively of truth-values that are falling *within* the second-level concept. Moreover, this parallel may be extended so as to also observe that physical objects are in the same way irrelevant for, and indifferent to the truth of the (second-order) proposition as irrelevant for, and indifferent to the truth-value of the proposition expressed by the main clause are the truth-values of the propositions that enter into the former as its clauses. This point is not however so transparent since it is completely muddled in Frege's definition of the conditional as a logical function "not peculiar to arithmetic" in which the sense of the peculiarity in question amounts to saying that this is a function that takes as its

[41] G. Frege, *Funktion und Begriff*, p. 24.

arguments any values such that these values are either the True or, if not the True, then it is the False *the value False including*, i.e., it is the False regardless of whether it is any object that is numerical or not, or the value False *per se*.[42]

Thus, it will be safe to say that all that the second-level concept in Frege's elucidations is about is the first-level concept alone, and thus whatever it says is *not* about *objects* as such not only in the formal sense in which Frege means it when he says that the concept word *men* stands for the concept itself but not for anything whatever that may appear as one of the objects (such as Chief Akpanya, for example) in the extension of this concept. For it happens to be true also quite literally, because all that saying that the first-level concept has instances means is that these instances are truth-values only, and not that these instances are instances of objects falling under the respective concepts—not even if the relevant truth-value happens to be the True. The thesis that whenever a concept has instances these instances are nothing but instances of objects falling under that concept would of course follow from Frege's Principle of Compositionality as its most immediate and direct corollary. But nothing can make it follow from instances to which Frege's Principle of Compositionality does not apply. That such instances exist could not be but immediately clear to Frege, who himself pointed out to clauses containing reported or indirect speech as clear exceptions. What Frege however ignored (perhaps merely failed to spot) is that the conditional in fact defied his Principle of Compositionality, since it exemplifies nothing other but yet another instance of an intensional context, to all instances of which Frege's Principle of Compositionality does not generally apply. Instance of an intensional context in the following sense: to the same extent to which in the context of belief that we posit as being true it is independent of the truth of what the belief is about, in the context of quantification in which we posit the conditional as being true the latter also turns up to be independent of whether its building blocks happen to be true or not and thus independent of whether they do or do not 'have instances'. In other words, in much the same way in which in *belief*-contexts what we are after and what we can talk about is the truth of the main verb but not the truth of the verb-complement, what we are after and what we can only talk about in a quantificational context is the truth of the conditional and,

[42] Cf. G. Frege, *Funktion und Begriff*, p. 28.

owing to this, only the truth-*value* of the functional expressions appearing as its antecedent and consequent. And, of course, speaking about the uniformity of the truth-values of functions representing concepts we are by no means speaking about functions (concepts) themselves, nor are we speaking about whether these concepts have instances or not to exactly the same degree to which by saying *John believes in unicorns* we do not assert anything about the existence of unicorns. Positing the quantifier in front of the respective functional expression is then very much like adding the definite article in front of *unicorns*: the resulting expression, *John believes in the unicorns,* will in the last analysis turn up to be only an abbreviation of what essentially is no more than *John believes in the unicorns in which he believes.*

Viewed against this background, the positing along with the True also of the truth-value False as the reference of a sentence serves, for Frege, a very specific purpose which appears of utmost importance for showing that arithmetic is reducible to logic. For the addition of the False as the referent of a sentence is connected with a very important aspect, or touch of sense, or rather foundational undertone in which „we see that in the reference of the sentence all that is specific is obliterated." What is specific that gets obliterated turns up to be also the specific status of Frege's logical function which, except for the expressly analytic case of '— is identical with itself', as in "Everything is identical with itself," does not always yield the truth-value True in the sense that it does not yield this truth-value under all substitutions of the variable for the values of the argument as the case happens to be with the algebraic equations in which the laws of arithmetic, such as $(a + b) \times c = a \times c + b \times c$, are expressed. It is exactly this sense in which Frege's logical function expresses just a particular case of the mathematical function in $f(x) = y$, but not in the least any generalisation over it.[43] Thus, the most that expressions like $f(x)$ and $f(x) = g(x)$, where the functional signs are conceived as logical and not as mathematical functions, can achieve is the status of algebraic equations like $x^2 = 1$ and $(x + 1)2 = 2(x + 1)$ which "always have the same value for the same argument, *viz.* the True for the arguments -1 and $+1$, and the False for all other arguments,"[44] and thus—as opposed to $(a + b) \times c = a \times c + b \times c$ which is true under *all* substitutions—are not lawlike expressions at all. It is exactly for this purpose, namely to represent

[43] Cf. above, p. 24.
[44] *Ibid.*, p. 16.

them so as they might be conceived as lawlike, that he puts the non-general logical function in $x^2 = 1$ and the equally non-general algebraic function in $(x+1)^2 = 2(x+1)$ in the form of an equation between the extensions of $x^2 = 1$ and $(x+1)^2 = 2(x+1)$ in

$$(4.25)\quad \grave{\varepsilon}(\varepsilon^2 = 1) = \grave{\alpha}((\alpha + 1)^2 = 2(\alpha + 1))$$

and then, reproducing $x^2 = 1$ and $(x + 1)^2 = 2(x + 1)$ in an assimilated form equally as concepts to the left side of the identity in his Axiom V:

$$(4.26)\quad (\grave{x} f(x) = \grave{y} g(y)) = \forall z(f(z) = g(z))$$

In what appears to the right side of the identity we can see the full sense of what he means by saying that

> If now the truth value of a sentence is its reference, then on the one hand all true sentences have the same reference and so, on the other hand, do all false sentences. From this we see that in the reference of the sentence all that is specific is obliterated.[45]

where what is specific that is obliterated is exactly the difference between the True and the False countenanced as the reference of statements rendered by means of functional expressions $f(z)$ and $g(z)$. For the full sense is that the identity in (4.26) to be true, it needs that the True and the False be both generalised as the reference of a sentence, and in this way as one and the same object that *falls* under the concept. Notice that in very much the same way that the left side expresses the identity between two different classes taken as referents of two different concepts,[46] what is expressed in the right side is the identity of two different truth-values taken as referents of the very same concepts now reconsidered as statements.[47] But what is asserted as being true by the use of the universal quantifier in the right side does not happen to be the functions $f(z)$ and $g(z)$ themselves, for due to the use of a different argument letter z its value is not restricted to those of x and y, and $f(z)$ and $g(z)$ may therefore both

[45] G. Frege. *Über Sinn und Bedeutung*, p. 35

[46] . . . which, as concepts, can be said to be true with respect to the objects that fall under them.

[47] . . . which, now as statements, will need to be said, because z is now a different letter than x or y appearing to the left, quite *regardless* of whether it is equal to y or x, to be true not only with respect to the objects that fall under the respective concepts and, thus, quite regardless of whether the object in question falls under the respective concept or not.

be either true or false. For even if the identity of x and y may be seen as being imposed by the very character of the identity in which they occur, this however gives no warrant that the value of z should, too, be identical, because the place where z occurs happens to be *outside* the scope of the identity in question, where such imposition can therefore not be a matter of fact. The general and, thus, analytic truth of the right side of the identity in (4.26) arises, then, as a synthetic identity holding between $f(z)$ and $g(z)$ as proper names due to both of them having the same reference, but this can be so only if the True and the False are entirely indistinct, i.e., regarded as one and the same truth-value, which is supported by the role of the implicit conditional that renders the truth-value true whenever both $f(z)$ and $g(z)$ have the same truth-value regardless of whether it is the True or the False. The proof of Frege's thesis (or rather its assumption in Axiom V) that the synthetic has its origin in the analytic and is thus *a priori* appears to be in the form of an analytic identity established between the two instances of synthetic truth to the left and to the right side of the identity in:

(4.27) $(f(z) = g(z)) = (f(z) = g(z))$

the True the False

where the truth of either of the instances of $f(z) = g(z)$ is synthetic due to the formal difference between $f(z)$ and $g(z)$, whereas the truth of the whole of (4.27) is analytic in virtue of the formal identity of the instances of $f(z) = g(z)$ appearing as one and the same proper name to the left and to the right side of the main sign of identity in (4.27). Thus, what the universal quantifier prefixed to the functional expression to the right side of the identity in (4.26) actually says is that, given the difference between the functions in $f(z)$ and $g(z)$, the value of the function is always identical to itself, in exactly the same way as the inscription in $\forall x(x = x)$ reads that every object is identical with itself. But in so far as what the identity in $x = x$ says is that the x to the left of the sign of identity has reference which is identical to that of the x to its right side, the identity in (4.27) tells us exactly the same, namely that the True is identical with the False. Thus, what we have in the quantifier expression on the right side is in deed the identity in

the True = the False

where 'the True' and 'the False' are the (different) proper names of one and the same object which is the reference of a sentence. But, since $\forall z(f(z) = g(z))$ is the right side of the identity in $(\dot{x}f(x) = \dot{y}g(y)) = \forall z(f(z) = g(z))$, then what the latter says is nothing else but that the identity of classes expressed in the left side of Frege's Axiom V is itself a contradiction. Thus we can see that Russell's paradox comes quite naturally as an inevitable corollary of what (4.26) presupposes, and already involves, as a premise. Thus, because Axiom V inherently contains the contradiction based on an implicit identity between the True and the False as the reference of universally quantified statements, the axiom gives warrant to any conclusion whatsoever—since what follows from a contradiction is literally everything—and, specifically, that the synthetic judgements of arithmetic represent knowledge that can be attained *a priori*, on the basis of logical truths alone.

2.5. *The Burden of Proof*

To understand why Frege still holds Axiom V as true despite the apparent air of paradox surrounding this main building block of the whole edifice of his system of logic, one needs to take a closer look at what might perhaps be termed as Frege's principle of the integrity of '*gleichbedeutend*' meaning. This seems to be the main principle on which Frege's belief in the analyticity of mathematical knowledge rests. And this principle shows up in Frege's notion of the "agreement in a certain respect in the form of an equation," that he uses in his letter to Peano in which he elucidates how different functions may be seen as identical in view of the identity of their value-ranges. This principle is, on the one hand, connected with Frege's understanding of identity in terms of identical 'characteristic marks' which he mentions in a letter to Liebmann:

> Very different from subsumption is the subordination of a first-level concept under a first-level concept: 'All squares are rectangles'. The *characteristic marks* of the superordinate concept (rectangle) are also *characteristic marks* of the subordinate one (square).[48]

and reiterates it, this time in terms of identical 'properties', in a letter to Peano:

[48] Letter to Liebman, in G. Frege, *Philosophical and Mathematical Correspondence*, p. 93.

> How, then, does 3 · 4 differ from 9 + 3? In nothing but the signs. There is no property which (3 · 4) has but (9 + 3) lacks, and conversely.[49]

And, on the other hand, this principle is linked with the way that Frege distinguishes between the principle of identity and its instance, which he touches upon in the context of his discussion of the identity of classes further on in his letter to Peano:

> Of course, we must then not place the equals sign between signs for equally long lines or equally large areas. But we can nevertheless make use of the equals sign even in these cases. Instead of putting the signs for the lines on both sides, we can . . . proceed as follows: every line determines a class of equally long lines; if two lines are equally long, the two classes coincide; and this again gives us an identity. Let *A* and *B* be lines and let '$A \cong B$' say that these lines are congruent. Then, in my notation, $\grave{\varepsilon}(\varepsilon \cong A)$ is the class of lines equal in length to *A*, and
>
> $$\grave{\varepsilon}(\varepsilon \cong A) = \grave{\varepsilon}(\varepsilon \cong B)$$
>
> is the essential content of '$A \cong B$' in the form of an equation. In similar cases, too, we can proceed in this way and express agreement in a certain respect in the form of an equation, without denying the meaning of the equals sign given above. Of course, this does not yet explain how it is possible that identity should have a higher cognitive value than a mere instance of the principle of identity. In the proposition, 'The evening star is the same as the evening star' we have only the latter; but in the proposition, 'The evening star is the same as the morning star' we have something more.[50]

Notice, however, that in the context of the principle of identity Frege also (and in the letter to Peano—primarily) speaks of definitions, and in the context of definitions Frege in turn speaks, in the *Grundgesetze*, of an introduction of a name by stipulating that it is to have the same sense and the same reference as some name composed of signs that are familiar. This seems to be equivalent with the case of Frege's examples of 3 · 4 and 9 + 3 above, which as he notices differ in nothing but the signs because there is no property which (3 + 4) has but (9 + 3) lacks, and conversely. And, since the characteristic marks of a concept under which an object falls are also the properties of this object, the identity of individual concepts (and thus senses) represented by 3 · 4 and 9 + 3 seems to be guaranteed by the identity of all the characteristic marks that appear to make the content of these concepts. But elsewhere Frege also speaks of the sense

---

[49] Letter to Peano, in G. Frege, *Philosophical and Mathematical Correspondence*, p. 127.

[50] *Ibid.*

of '5 + 2' as different from the sense of the combination of expressions '4 + 3'.[51] This contradiction is however unavoidable since what Frege appears to be speaking about is that he regards the recognition of one class as identical with another as more valuable than a mere recognition that every class is identical with itself in much the same way as it happens when we are concerned with individual objects. And in this way the introduction of a truth-value as the reference of a sentence serves the same role for establishing the referential identity of two common names or, as Frege calls them, concept-names as the heavenly body under the name of *Venere* which is in outer space and which is known to have mass serves as a common reference of the two proper names, *Lucifero* and *Espero*, for the recognition of their identity in *Il Lucifero è identico col Espero*.

The respect in which Frege regards the difference between an instance of the principle of identity and the principle of identity itself is exactly the respect in which *Il Lucifero è identico col Espero* is different from a proposition in which *concept-names* that stand for concepts appear in the places which, in *Il Lucifero è identico col Espero*, are occupied by *proper* names *Lucifero* and *Espero* that stand for individual objects alone. In other words, the whole difference lies in the distinction between proper names and concept-names, the latter being the bearers of the principle of identity whereas the former seem to exemplify only its instance having to do with objects for which proper names stand. But the kind of relationship in these cases is entirely different: what we have in the first case is a first level relation between physical objects, and in the second case—a second level relation between concepts, "which corresponds to, but should not be confused with, equality (complete coincidence) between objects."[52] Because, contrary to objects, concepts are not exactly objects—they are objects of an entirely different kind: they are *logical*, not physical objects, and thus objects only in a metaphorical sense. Accordingly, their relationship cannot be mediated by Relation I, that of an object being subsumed by a first-level concept under which the object on the other side of the identity also falls, but by Relation II which is rendered by the conditional. The relationship between the truth-values that appear in the place of objects as the corresponding reference will then need to be fashioned in exactly the same way as the respective relationship between the identical objects: the

[51] *Ibid.*, p. 128.
[52] G. Frege, Comments on Sense and Meaning, in *Posthumous Writings*, p. 121.

identity in question would hold if the truth-values in question are one and the same truth-value, otherwise not. But the truth-value of the conditional must therefore always be the truth-value True, for otherwise it would simply mean that the identity does not hold because the truth-values of the respective concepts do not coincide. Thus, in contradistinction to what we have in the case of a mere instance of the principle of identity, where the material necessary condition for the identity to hold is in that the object in question must necessarily fall under the respective concept, the principle of identity as such appears to be independent of the requirement that the objects necessarily fall under the concepts. It appears to depend on the issue of the *simultaneity* of what the objects *do*, and thus quite irrespectively of what they *actually* do, i.e. regardless of their falling or not falling under respective concepts. In this way, the issue of whether the quantifier actually allows us to pick an x or not and thus whether quantification exhibits the choice-like nature shows itself as quite irrelevant to the issue of whether mathematics can indeed be reduced to logic in much the same way as it shows itself irrelevant to Frege's Axiom V.

Now, if a proper name that stands for an object expresses sense as well as has this object as its reference, what concept-names stand for are not objects but concepts. But truth-value is not part of thought, thus nor is it part of sense and—respectively—nor is it part of concept. Therefore, what a concept-name has as its reference is not a truth-value, for the latter is not part of what it stands for. A concept is basically unsaturated and it receives a truth-value only if a proper name occurs as a part of an expression, without which we would be dealing with an unsaturated expression standing for a concept. But expressions such as *man* or *mortal* are syntactically primitive and, unlike such expressions as '( ) is a morning star', they cannot be 'saturated' by proper names. Instead, they (or rather the concepts that they stand for) are saturated directly by all objects that belong to the value-range of a respective function and thus make the concept saturated. According to Frege, this is exactly the case when 'saturation' is not purely arbitrary and thus what is construed in terms of function and argument acquires a *substantive* {*inhaltlich*} significance. 'Saturation' thus appears to be exclusively a matter of just two instances of the occurrence of the concept-name in a predicative context, i.e. either in the context of Relation I where the use of deictic expressions and proper names eliminates the occurrence of quantifier-words, e.g. *This is a man* (the only exception being *I'm every woman*), or else in the context of quantification that is strictly confined to the relation of subordination, Relation II, that involves concepts alone. But in the latter, any attempt to 'saturate' the respective concepts automatically transforms the respective concept-words into names proper, i.e., proper names that stand directly for objects and not, as concept-names, for concepts. Moreover, it also automatically transforms the second level relation between concepts into a first level relation between objects, which in itself is a contradiction because concepts are not objects and whatever comparison of them as objects will not be their comparison as concepts and thus the very point of comparison will be missed.

In order to resolve this, Frege looks not for identities but for analogies that could be drawn between the properties of equality in the case of objects, on the one hand, and concepts, on the other. Looking for these analogies or, as Frege calls them—strong affinities with the first level relation of equality (identity), he notes that:

> The properties of equality, that is, which we express in the sentences '$a = a$', 'if $a = b$, then $b = a$', . . . have their analogies for the case of that second level relation. And this compels us almost ineluctably to transform a sentence in which mutual subordination is asserted of concepts into a sentence expressing an equality.[53]

What this basically means is that a sentence expressing an equality of concepts is that in which what is asserted of concepts is the fact that their subordination is mutual. An example of such mutual subordination of concepts Frege gives in (5.28):

$$(5.28) \quad (a^2 = 1) \overset{a}{\approx} ((a + 1)^2 = 2(a + 1))$$

However, the mutual subordination between $(a^2 = 1)$ and $((a + 1)^2 = 2(a + 1))$ that we find in (5.28) will not suffice if we want to express the equality of these logical functions *resp.* concepts, because "[w]hat we have here is that second level relation which corresponds to, but should not be confused with, equality (complete coincidence) between objects."[54] Indeed, what Frege uses as a proxy sign of equality plays here the same role as the same kind of a proxy sign of equality would play in a first-level relation of identity in:

$$(5.29) \quad (a \text{ is the morning star}) \overset{a}{\approx} (a \text{ is the evening star})$$

---

[53] G. Frege, On Schoenflies: Die Logischen Paradoxien der Mengenlehre, in *Posthumous Writings*, p. 182.

[54] G. Frege, Comments on Sense and Meaning, in *Posthumous Writings*, p. 121.

where it would state no more than that both occurrences of a in (5.29) are coreferential. But to know that a would indeed be the same object even though disguised under two different names we must first discover that it is not the fact—in the case when a = *Venus*—that the statement 'a is the morning star' is true while the statement 'a is the evening star' is false or else that 'a is the evening star' is true whereas 'a is the morning star', false. In other words, in order to know that the assumed identity at Level I holds, we need first to enquire if the respective object that is an entity of the first level enters into the same relationship of *falling under* with each of the entities of Level II and, thus, we need first to be aware of whether *Venus is the morning star* as well as *Venus is the evening star* are both true. Then, by mere analogy, what seems to warrant that the same type of relationship as that of identity between objects holds also in the case of concepts is that the same scheme should also apply at the level of concepts, so that the concepts in question would be seen as entering into the same kind of relationship with certain entities that play with respect to them the same role as concepts themselves play with respect to objects. What we need, therefore, is a concept of level-(n+1), which will play the same role with respect to concepts of level-n as the latter themselves play with regard to objects. Then, what we will express with the help of Frege's quantifier, introduced as a required concept of level-(n+1), in

$$(5.30) \quad \vdash\!\!\smile^{\mathfrak{a}}\!\!- (\mathfrak{a}^2 = 1) = ((\mathfrak{a} + 1)^2 = 2(\mathfrak{a} + 1))$$

will be precisely the principle of identity as such, of which the first-level identity in $a = b$ will be seen as an instantiation if we represent correlations found between levels n and n+1 as having been "moved down a level."[55]

But, considered in the same terms as the identity at Level I, the identity at Level II should then be re-written as

$$(5.31) \quad \vdash\!\!\smile^{\mathfrak{a}}\!\!- f(\mathfrak{a}) = g(\mathfrak{a})$$

where the signs of equality in both $(a^2 = 1)$ and $((a + 1)^2 = 2(a + 1))$ are eliminated by replacing either of these expressions by the signs for respective logical functions that are deemed identical, which entitles us to handle the numerical identities in $(a^2 = 1)$ and $((a + 1)^2 = 2(a + 1))$ as cases of a mere subsumption of the values of a under the respective concepts now

[55] G. Frege, On Schoenflies: Die Logischen Paradoxien der Mengenlehre, in *Posthumous Writings*, p. 182.

re-written as '$f(\ )$' and '$g(\ )$'. But by the same token, however, we will also be forced to eliminate the sign of identity in (5.31)[56] in favour of the identity between the two conditionals, as in:

$$(5.32)\quad \vdash\!\!\overset{a}{\smile}\!\!- f(a) \supset g(a) = g(a) \supset f(a)$$

which by itself indicates nothing more and nothing else but that in (5.31) we were dealing not with the sign of identity proper, but with the sign of the same kind of proxy identity as in (5.28). But in order to clarify the meaning of the sign of equality in (5.32), we have no way out but to conceive (5.32) as itself being a case of mutual subordination, this time as a case of the mutual subordination between $f(a) \supset g(a)$ and $g(a) \supset f(a)$now appearing as complex concepts standing in place of simple concepts, $f(a)$ and $g(a)$, in (5.31), which we will then have to represent as something like (3.33):

$$(5.33)\quad \vdash\!\!\overset{a}{\smile}\!\!- (((f(a) \supset g(a)) \supset (g(a) \supset f(a)) = ((g(a) \supset f(a)) \supset ((f(a) \supset g(a)))$$

But in this case it will immediately be evident that the 'meta'-sign of identity in (5.33), which we will need as an explanation (or else justification) of the proxy sign of identity in (5.32), will readily be seen as itself standing in need of explanation. Because Frege's stipulation of the new function for the sign of identity—that of the identity between the truth-values as referents of the functional expressions—does not change the fact that what we have as the result of this transformation is none other but mutual subordination of concepts only transferred one level up, and which is still there regardless of whether we introduce the quantifier expression to the left of the equation or not. Then, since in (5.33) the sign of identity, too, will remain unexplained in exactly the same way as the sign of identity which it itself explains, the procedure will have to be repeated, and this will have to be done so on, and so forth, *ad infinitum*.

Notably, with respect to his Axiom V, Frege himself asks a question:

> But by what right does such a transformation take place, in which concepts correspond to extensions of concepts, mutual subordination to equality? An actual proof can scarcely be furnished.[57]

---

[56] ... which is now not a sign of the identity of Level I, as in '$a = b$', but a sign standing for a biconditional expressing the "mutual" character of concept subordination.

[57] G. Frege, On Schoenflies: Die Logischen Paradoxien der Mengenlehre, in *Posthumous Writings*, p. 182.

as well as provides the following answer:

> We will have to assume an unprovable law here. Of course it isn't as self evident as one would wish for a law of logic. And if it was possible for there to be doubts previously, these doubts have been reinforced by the shock the law has sustained from Russell's paradox.[58]

Indeed, Frege's wording here does seem to indicate that he is far from being aware that an actual proof, if attempted, would inevitably generate infinite regress. Because what he wants to state as a law in step II of his construction of Axiom V, namely the equivalence of mathematical and logical functions,[59] is already contained in step I as a premise, where the particular case of a mathematical function $f(x) = y$ in $a^2 = 1$ is taken as equivalent to a particular case of the function of a logical kind, '$f(a)$ = truth-value True'. For it is exactly his stipulation of truth-value as reference that inevitably leads to infinite regress in case we were to take the burden of any such 'actual' proof. And it is exactly for this reason that the conception of the quantifier as a second-order function taking truth-value True as its argument turns up to be demonstrably false. As a corollary, since it is nothing but this stipulation alone on which the whole idea of variable-binding quantifiers rests, its further elaboration in terms of perfect *vs.* non-perfect information strategies happens to be nothing but a false currency that has only imaginary, but absolutely no factual, genuine backing in the logical reality of quantification.

[58] *Ibid.*

[59] ... specifically, the equivalence between an algebraic equality $a = a$ where a denotes a numerical object (and expresses sense) and a logical equivalence $p = p$ where p denotes a truth-value (and expresses a thought).

THE TRANSCENDENTAL ANALYTIC OF QUANTIFICATION

CHAPTER THREE

REFERENCE AND CAUSALITY

> Apparent discrepancies between natural language and formal logic provide no evidence of any defect in natural language: they are rather evidence that our analysis of natural language is deficient, or our formalization of logic is deficient, or our understanding of the relationship between language and logic is deficient, or our data reflect the interaction of language and logic with some third factor that we have not yet properly accounted for.
>
> *James D. McCawley, Everything That Linguists Have Always Wanted to Know about Logic, pp. xix–xx*

> 'Well! I've often seen a cat without a grin,' thought Alice; 'but a grin without a cat! It's the most curious thing I ever saw in my life!'
>
> *Lewis Carroll, Alice's Adventures in Wonderland, Ch. VI, Pig and Pepper.*

3.1. *'Hilfssprache' vs. 'Darlegungssprache'*

As has been shown in the foregoing account, Frege's reservations as to whether a single concept-word like *men* preceded by a quantifier happens to have any real man as its reference appear to be applicable—generally—to all expressions standing for the so-called complex concepts such as *if-man-then-mortal*. For the only thing that Frege assumes to be only able to make the latter referential is his expression of generality prefixed to the conditional which makes the logical skeleton of *if-man-then-mortal* as a complex concept. But, in the first place, the conditional—along with the quantifier itself—is nothing but just another case, apart from the constant/variable distinction as such, showing how a misguided extension of a mathematical notion can go wrong. As a matter of fact, the only referentiality of the conditional turns out to be its purely truth-functional referentiality in the sense of reference to truth-values only. Russell was right in pointing out that there is nothing about *all Greeks* in *if-anything-is-a-Greek-it-is-a-man*, for all that there is is about truth-values only.[1] Evidently, from this it does not follow at all that the quantifier, being itself dependent on the truth-value readings of the conditional based on the truth-value readings of Ψ and Ξ, imposes—by virtue of this dependence—whatever referential readings on Ψ and Ξ themselves and thus tells that what Ψ and Ξ are about is everything in the world. As a corollary, Frege's quantifier does not say anything about whatever objects in the world not only in the case when Ψ and Ξ appear singly, but also when they are combined by means of the conditional. It is simply that Frege's quantifier has no ability of doing this: for if it had, then it would mean the annihilation of the conditional as such. As has been shown in Section 1.9, this annihilation is immediately evident at the very moment that we start assuming that there indeed *is* some object x about which the quantifier asserts something (literarily, that x exists). And if we have to admit this, then we will also have to admit that Frege's doctrine of the conditional combined with his expression of generality intended as a means of making overt the causal connection implicit in the word "if" appearing in place of the former subject-predicate distinction[2] happens to be fundamentally flawed. Along with this, the whole of the alleged breakthrough of modern logic resting wholly upon the rejection of a basic Aristotelian idea, namely, that sentences have subject-predicate form,[3] turns in the last analysis to be built on philosophical illusion.

The roots of this philosophical illusion can be traced to Frege misapprehension of the thesis of Trendelenburg concerning the psychological increment which allegedly prevents ordinary language to fully realise the task of the linguistic sign and makes it imperfect and, thus, prevents the linguistic sign of being a genuine sign—and, thus, directly to Frege's antipsychologism. Assuming for the moment that the reflection on causality makes the centrefold of the antipsychologistic essence of Frege's effort, it will not be hard to see that Frege's antipsychologism misses an essential point he is trying to make. His basic idea which he endorses by his vast anti-psychologistic rhetoric amounts to saying that the grammar of natural language obscures and distorts—in effect defeats—the cause-

[1] Cf.: "The statement 'all Greeks are men' is thus much more complex in form than the statement 'Socrates is a man'. 'Socrates is a man' has 'Socrates' for its subject, but 'all Greeks are men' does not have 'all Greeks' for its subject, for there is nothing about 'all Greeks' either in the statement 'there are Greeks' or in the statement 'if anything is a Greek it is a man'." (B. Russell, *History of Western Philosophy*, p. 219).

[2] *Begriffsschrift*, §5.

[3] Cf. Dag Westerståhl, *Quantifiers in Formal and Natural Languages*, p. 1.

result relationships that arise between the signs and the concepts that they stand for, and this is exactly that which prevents them from being genuine signs. What is nevertheless left by his analysis as no less obscure is the exact reason for which the signs of *his* concept-script should be seen as genuine signs, and thus signs of a language characterised by a fully internal—and thus strictly causal—relation between the sign and the content of the signified concept. For what he in effect achieves by his demonstrations is an artificial language which, as he believes, matches or at least is able to match exactly the cause-effect relationships. But the internal, or cause-effect relations that he talks about happen to be those arising between concepts themselves. These, however, are not the relations between concepts and signs so that the latter could be considered, in contradistinction to natural language, as signs of a language that brings 'the shape of the sign in direct contact with the content of the concept'. For nowhere does he give any account of why the allegedly causal, internal relations between the concepts should also count, at the same time, as causal relations between signs and the concepts that they stand for. Quite obviously, the latter do not follow from the former. The only possible explanation that we could admit here is that we construe these relations as rigid designation as on Kripke's causal theory of names. But in this case, too, we would have to start with some initial "baptism" or dubbing of the concepts with their respective signs, and it would not slip unnoticed that in this case, too, the signs for concepts cannot be "discovered" but only "invented" and, thus, no one will be in a position to assert that these signs existed prior to their putative discovery. For, even if we claim that the causal connection might have existed prior to our "discovery" of an appropriate sign, then this causal connection would not be direct—as Frege claims—but necessarily mediated by mind. Quite possibly, an erring one. But even if the mind in question turns eventually to be right as to the presence of a causal connection, this connection will never be such as suggested by Frege, *viz.* that of signs standing *directly* for thoughts. For this would necessarily involve the presence of mind involved in the process of dubbing and thus the very necessity of the act of initial dubbing already reveals the presence of a purely subjective component in any "direct" link between the sign and the concept that it stands for, and by this also the presence of an indispensable linking chain going through the mind of the "discoverer" of such causal connection.

Thus, despite Frege's antipsychologistic argument, his logic turns up to be neither free from subjectivity or psychological adulteration, nor does it lead him out of 'subjective' psychology even into the realm of Kant's 'transcendental' psychology. In this way, the whole idea of Frege's to construe some sort of mind-independent logic shows itself as demonstrably false. For what appears, in this reasoning, to be of no concern for logic is not only the mystery of how it is possible that human minds grasp mind-independent thoughts since they are imperceptible, but even to the more extent the mystery of how these human minds can "discover" the immediate connection between imperceptible mind-independent thoughts and perceptible mind-dependent symbols of quantificational logic. Or should we also claim that the symbolic artefacts of quantificational logic are—owing to the inner connection that they have with thought—in deed part of the mysteriously grasped third-realm entities, and thus they are directly grasped in exactly the same way as thoughts that they express? In any case, however, Frege's idea of mind-independent logic shows itself as unsustainable the very moment that we resort to any sort of baptism, which immediately takes us back to the sphere of the psychological.

In effect, Frege himself grossly compromises his own antipsychologistic platform when he states in one of his later writings that

> Once we have come to an understanding about what happens at the linguistic level, we may find it easier to go on and apply what we have understood to what holds at the level of thought—to what is mirrored in language.[4]

Because, as he concedes,

> Language may appear to offer a way out, for, on the one hand, its sentences can be perceived by the senses, and, on the other, they express thoughts.[5]

However, whatever it is supposed to be perceived by senses happens to be not in the least the inner connection between the content of the signified idea and its sign in Frege's *Begriffsschrift*. Moreover, Frege's logical notation appears to be designed not for the proper reflection of thoughts but only for the reflection of their generality:

> Here it isn't a question of the day to day understanding of language, of grasping the thoughts expressed in it: it's a question of grasping the property of thoughts that I call generality.

---

[4] G. Frege, 'Logical Generality', in *Posthumous Writings*, p. 259.

[5] *Ibid.*

But, as he concludes,

> Of course for this we have to reckon upon a meeting of minds between ourselves and others, and here we may be disappointed.[6]

By these words Frege himself seems to be indicating the limits of his antipsychologistic conception. And, most importantly, by saying this he also admits that these limits are actually, and most conspicuously, cutting across the central issue of Frege's concern, which boils down—in his own terms—to the "question of grasping the property of thought that I call generality," and for which, as he himself recognizes readily, "we have to reckon upon a meeting of minds between ourselves and others."

What Frege, by stating this, in fact actually recognizes is that his concept-script appears to be in the same kind of connection with the concepts it is intended to express as natural language itself, and in practice is no more than simply a kind of shortcut to thought, whose connection with it is necessarily mediated by, and in fact established through, natural language as such. Because, as Frege explains, in using his notation which he calls '*Hilfssprache*' as an intended shortcut to thought:

> ... we are stepping outside the confines of a spoken language designed to be heard and moving into the region of a written or printed language designed for the eye. A sentence which an author writes down is primarily a direction for forming a spoken sentence in a language whose sequences of sounds serve as signs for expressing a sense. So *at first there is only a mediated connection* set up between written signs and a sense that is expressed. But once this connection is established, we may also regard the written or printed sentence as an immediate expression of thought, and so as a sentence in the strict sense of the word.[7]

If we represent the relation between Frege's *Hilfssprache* or *written language* ①, *spoken language* ②, and *sense* ③ diagrammatically, as in (1.1):

(1.1)

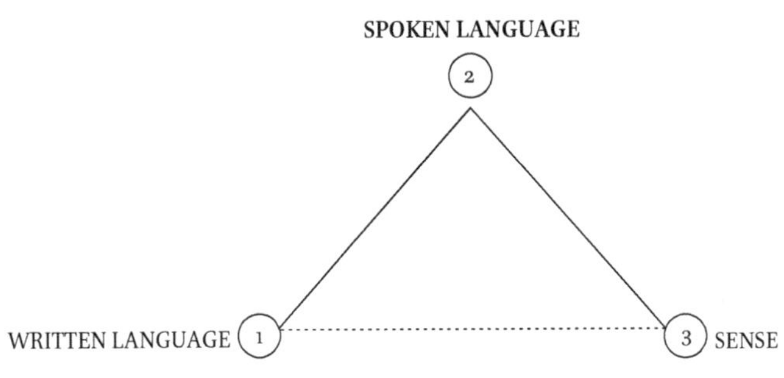

[6] *Ibid.*
[7] *Ibid.*, p. 260.

it will be seen that it appears to be the same as in the semiotic triangle:[8]

(1.12)

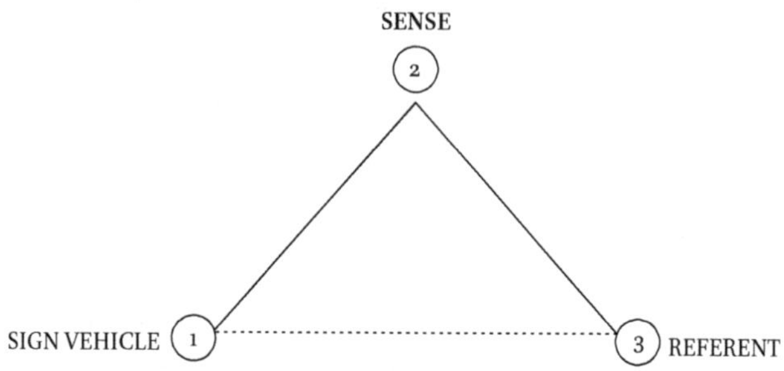

where the dotted base line indicates the indirect nature of the relationship between the sign vehicle and the referent and thus the path of mediation from (1) to (3). Thus, the direct nature of the connection between Frege's concept-script and thought turns up highly suspect and, in the last analysis, unsustainable due to the mediating effect of human speech and, thus, the inescapable presence of mind owing to which "this connection is *established*." In this way, the "sentence in the strict sense of the word" of Frege's notation appears to be riddled with psychologistic interpretation to a no less extent than a corresponding sentence in, say, ordinary German which Frege terms as '*Darlegungssprache*'. And since, according to Frege, for what we have in his words "to reckon upon a meeting of minds" is not only "a question of the day to day understanding of language" and thus "of grasping the thoughts expressed in it" but also a question of grasping their property, i.e. generality, with which his *Hilfssprache* is expressly concerned—they both seem to be equally exposed to the same "deteriorating" influence of a human factor.

However, it is not exactly *this* kind of deficiency that really bothers Frege, for the deficiency that Frege is really concerned with appears to be in a very distant connection with the alleged psychological "adulteration" of the grammar of natural language. For the *real* source of its deficiency he sees not so much in the fact that sentences in natural language have subject-predicate form,[9] but in (what he assumes as an established fact) that

[8] W. Nöth, *Handbook of Semiotics*, p. 89.

[9] ...which, according to his own statement, he did not entirely exclude in his initial considerations on *Begriffsschrift*, so his refutation of Aristotelian categories was not a mere consequence of the principle of sharp separation of the psychological from the logical but, first and foremost, it was a necessary step in eliminating the main obstacle that he saw

natural language does not possess the necessary means for the expression of generality. And the real *issue* Frege sees not in the devastating effects of the psychological or neurological underpinnings of our linguistic ability, nor does he see it in whatever properties of spoken language that would facilitate the connection in question, as the property of the linguistic sign that Frege seeks to expose in his *Hilfssprache* happens to be the one that spoken language, taken as an intermediary link between his notation and thought that it expresses, does not possess at all. For the most important and, indeed, the most fundamental question that informs all of his logico-linguistic inquiry and which Frege directly addresses is the one that he puts forward in his essay on *Logical Generality*: "Now, in what form does generality make its appearance in language?"[10]

It would be hard, however, to find cohesion or consistency in the way Frege provides an answer: without resolving the main question of philosophical grammar as it was formulated by Trendelenburg, Frege's attempt remains more of a piece of guesswork for which he merely anticipates confirmation from philosophical quarters,[11] being rather sure that his intuitive guess is as flawless and unmistakable as "grasping" of a "true" thought can only be. The major inconsistency reveals itself on comparing Frege's question above with his statement that

> Language may appear to offer a way out, for, on the one hand, its sentences can be perceived by the senses, and, on the other, they express thoughts. As a vehicle for the expression of thoughts, language must model itself upon what happens at the level of thought. So we may hope that we can use it as a bridge from the perceptible to the imperceptible. Once we have come to an understanding about what happens at the linguistic level, we may find it easier to go on and apply what we have understood to what holds at the level of thought—to what is mirrored in language.[12]

On a careful reading of this passage it will be readily seen that the real question that Frege deals, or rather is attempting to deal with is not exactly the question that he makes the impression to be asking, by saying

---

as standing in the way of finding what proves to be a convenient device for talking about generality.

[10] *Ibid.*, p. 259.

[11] Cf. Frege's 'Über die wissenschaftliche Berechtigung einer Begriffsschrift', pp. 113–14:
The use of my symbols in other areas is not excluded through this.... Whether this happens or not, the intuitive representation of the forms of thought has in any case a significance that goes beyond the area of mathematics. For that reason, *may philosophers pay some attention to this matter.*

[12] G. Frege, 'Logical Generality', in *Posthumous Writings*, p. 259.

"Now, in what form does generality make its appearance in language?" For the real question is, in deed, quite different: what Frege means by asking this question is "How do we understand or, rather, how do we perceive generality which is imperceptible otherwise but at the linguistic level so that we could go on and apply what we have understood to what holds at the level of thought—to what is mirrored in language?" In other words, the question that Frege *really* addresses is "How should we interpret the appearance of generality in '*Darlegungssprache*' (ordinary German) so that we could apply it to the representation of generality in '*Hilfssprache*' (his logical notation) and, thus, not "In what form *does* generality make its appearance in language?" but, rather, "In what form *should* generality make its appearance in language?"

Thus, Frege's implicitly targeted question "In what form *should* generality make its appearance in language?" partly presupposes the answer as to the form in which generality *makes* its appearance in language. The second inconsistency in Frege's argument that therefore directly follows from the first is in that whatever he sees in the expression of generality in language is not that how it in fact reveals itself in natural language, but how it *should* reveal itself were it to "mirror" the thought correctly. Of course, in this case "correctly" means not being in strict accordance with linguistic facts connected with the expression of generality, but in accordance with his prior assumptions as to how perceptible linguistic facts would look like should they be in accordance with imperceptible facts of thought prior to its expression by means of perceptible signs.

And, indeed, Frege's answer to the question that he poses in his essay on *Logical Generality* turns up to be, in effect, not about the form in which generality makes its appearance in language but, rather, about the form in which it *should not* do so. Of the three examples that he adduces, i.e.:

'All men are mortal'
'Every man is mortal'
'If something is a man, it is mortal'

Frege finds the first two unsuitable for logical purposes for the reason of their being "not suitable for use *wherever* generality is present, since not every law can be cast in this form."[13] The real reason lies, of course, in the fact that the first two examples are not rendered in the form of the conditional. The third example, therefore, serves for confirming this thesis since

[13] *Ibid.*

it is rendered in the form of the hypothetical compound sentence. But even this example Frege discards as equally unsuitable for use *wherever* generality is present since in this case, too, "we would only be able to deal with the very simplest cases." To avoid the erring effect of using words like 'something' and 'it' in the third example, which Frege finds unworthy of consideration since 'something' and 'it' as expressions of generality are clearly not equiform, he takes a corresponding hypothetical compound sentence, 'If Napoleon is a man, Napoleon is mortal', in which the condition of equiformity is being met by means of replacing what he calls "the indefinitely indicating parts of the sentence" by one and the same singular term, 'Napoleon'. The resulting sentence, however, is a singular statement, which therefore does not qualify for the expression of generality, even despite its being dressed in the form of a hypothetical sentence.

This singular statement is nevertheless regarded as a conclusion, and thus as a particular case of 'If *a* is a man, *a* is mortal' obtained by substituting for the indefinitely indicating letters equiform with '*a*', the proper names of the form 'Napoleon' by freeing the two sentences from quotation marks, thus making it possible to put them forward with assertoric force. Presumably, since 'If *a* is a man, *a* is mortal' expresses the same thought as 'If something is a man, it is mortal' put equally in the form of a general statement, then the substitution of the proper names of the form 'Napoleon' for the indefinitely indicating letters equiform with '*a*' warrants the transition from the general statement in 'If something is a man, it is mortal' = 'If *a* is a man, *a* is mortal' to a corresponding singular statement in 'If Napoleon is a man, Napoleon is mortal', which must also be true if 'If something is a man, it is mortal' expresses a true thought.[14] But, as a presumption, it is immediately faced with a problem. The problem is, however, that even though *If something is a man, it is mortal* happens to be a general statement (which is either universal or particular), Frege's use of *something* in a general statement in *If anything is a man, it is mortal* does not necessarily make it universal and, thus, the latter does not necessarily have to be one expressing the form of a general law. As a matter of fact, the general statement in question happens to acquire not the form of a universal statement, but that of a *particular* statement, so the inference in question may in this case only proceed from the singular statement to the particular one, but *not* vice versa.

[14] Cf. also D. Wiggins, *'Most' and 'All': Some Comments on a Familiar Programme, and on the Logical Form of Quantified Sentences*, pp. 330 and 345 n 20.

The curious thing is that Frege, quite evidently, treats the three sentences above as different expressions of one and the same thought, since that remark that he immediately adduces to these examples is: "[t]he differences in the expression do not affect the thought itself." Elsewhere, he also makes it explicit that:

> Language has the means of letting now this, now that, part of the thought appear as subject. One of the best known is the distinction of active and passive forms. It is therefore not impossible that the same thought appears in *one* analysis as singular, in another as existential, and in a third as universal.[15]

Quite possibly, the key to a proper understanding of this may be found in the *Begriffsschrift*, where Frege makes it clear that:

> ... the different ways in which the same conceptual content can be considered as a function of this or that argument have no importance ... the whole splits up into *function* and *argument* according to its own content, and not just according to our way of looking at it.[16]

Presumably, Frege finds in his examples above exactly the case in which we only have such different linguistic ways of dressing the same thought which make it *appear* in one analysis as singular, in another as existential, and in a third as universal. For Frege, it is exactly the same thought which is revealed as truly universal only when split up into function and argument with the use of ordinary words on the one hand and (uniformly) variable letters on the other. Any kind of substitution will then only make it appear as singular (as a result of the use of proper names instead of letters) or else existential (as a result of the reverse substitution of the 'indefinitely indicating parts of the sentence' for the letters themselves), but the thought in both cases will remain essentially the same insofar as the letters appearing in the antecedent and the consequent of the conditional are uniform with one and the same letter and, thus, insofar as what we use as letters in the antecedent and the consequent of the conditional are nothing else but tokens of one and the same letter as their type. In this case, the substitution of these token letters by the respective tokens of one and the same proper name will give us *something* like a singular case, and by the 'indefinitely indicating parts of the sentence' like 'something' and

[15] G. Frege, *Kleine Schriften*, p. 173. English translation in H. Sluga, *Gottlob Frege*, p. 136.

[16] *Begriffsschrift*, § 9.

'it'—*even though the latter quite "occasionally" happen to be non-uniform*—a particular case or else instance of the same universal thought in 'If *a* is a man, *a* is mortal'.

The problem that Frege faces does not, however, boil down entirely to the lack of the required kind of equiformity in ordinary language that his examples of expressions of generality clearly show. Obviously, ordinary language cannot provide examples that Frege needs for his expository purposes since, on the one hand, placing the 'it' in both the antecedent and the consequent of the conditional readily exposes the *non sequitur*—as the conditional in 'If *Napoleon* is a man, *Napoleon* is mortal' clearly does not follow from 'If *it* is a man, *it* is mortal'. And, on the other hand, the equivalent use of 'something' with respect to equiformity as in 'If *something* is a man, *something* is mortal' leads immediately to the *non sequitur* in the conditional itself. Moreover, even the use of indexicals proper such as 'it' or 'this', as in 'If *it* is a man, *it* is mortal', is easily defeated by an example like Wittgenstein's "This is beautiful and this is not beautiful" (pointing at different objects)[17] which in theory, according to Wittgenstein, should pose a contradiction, though in practice it does not. Evidently, Frege, in much the same manner as later Wittgenstein, treats such kind of content variability in the case of indexicality, i.e. *monosemous* content variability,[18] as just one of the instances of the assumed logical defectiveness of ordinary language. The problem lies much deeper, however, since exactly the same kind of defectiveness, insofar as the phenomenon of indexicality is concerned, affects in the very same manner not only '*Darlegungssprache*' (ordinary German or, say, English), but also the representation of generality in '*Hilfssprache*' (his logical notation), too.

3.2. *Frege's Constant/Variable Distinction vs. Peirce's Type/Token Distinction*

In his essay on *Logical Generality* Frege clearly construes his sentence, 'If *a* is a man, *a* is mortal', as a semantically correct model for which he finds it natural to copy the methods of arithmetic. In this setting, the corresponding sentence, 'If something is a man, it is mortal', plays no longer

[17] L. Wittgenstein, *Remarks on the Philosophy of Psychology*, § 37.
[18] Cf. M. Pelczar, *Wittgensteinian Semantics*, pp. 486–87.

the role of a direction for forming a spoken sentence—rather, it becomes a derivative, merely written copy of the spoken sentence as such:

(2.1)

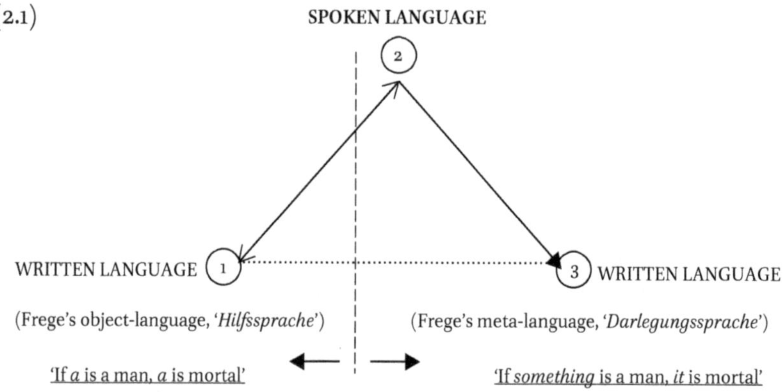

Given in very broad terms that the object-language appearing on the left side of the diagram is, according to Frege, to serve for us as a bridge from the perceptible (signs) to the imperceptible (thought), the specific task of this device—in Frege's terms—is narrowed to that of grasping, and thus making perceptible through visible signs, solely the property of thoughts that he calls logical generality. This specific goal seems to have been obtained by the implementation of Frege's constant/variable distinction, due to which the expression in Frege's object language is seen as construed of "two different *constituents*: *those with the form of words and the* individual *letters*".[19] But, since "[t]he former correspond to words of the spoken language," whereas "the latter have an indefinitely indicating role,"[20] all the gain in attaining perspicuity in expressing logical generality through the signs of Frege's object language appears to have been obtained by the mere replacement of the indexicals like 'something' and 'it' by variables or, in Frege's terminology, "by selecting letters for indefinitely indicating parts of a sentence."[21] The advantage of the resulting expression in Frege's object-language, 'If *a* is a man, *a* is mortal', lies in Frege's view in the fact that:

---

[19] G. Frege, *Logical Generality*, p. 260.

[20] *Ibid.*

[21] *Ibid.*

> Here the equiform letters cross-refer to one another. Instead of letters equiform with '*a*' we could just as well take ones equiform with '*b*' or '*c*'. But it is essential that they should be equiform.[22]

As can be easily seen from the citation above, the whole purpose of the implementation of the constant/variable distinction boils down to the elimination of ordinary words, corresponding to 'indefinitely indicating parts of a sentence', by replacing all their occurrences in a given sentence with what in Peirce's terminology would be called tokens of one and the same letter taken as their type, irrespectively of whether it would be '*a*', '*b*', or '*c*', etc. As a result, the whole distinction appears as a distinction between signs which are words of the spoken language, on the one hand, and tokens of some arbitrarily chosen letter taken as their type, on the other. And what Frege takes as essential is that the tokens in question should cross-refer to one another. Notably, this time their assumed cross-reference to one another is not introduced by mere convention, as in the introductory paragraph of his *Begriffsschrift*, but explained by cross-reference which according to Frege comes out as admittedly necessary, unavoidable, and indispensable due to their being token letters of one and the same letter as their type.

How, then, does the type/token distinction apply to the level of thought? If it does, what can it reveal insofar as the essence of thought might be concerned? For, if the equiform instances of '*a*' are tokens, they clearly *cannot* be linked with generality. How can therefore their meaning be general if their content is *indexical*?

The answer to the last question may be sought in the fact that the differentiation between constant meaning and variable content[23] applies to indexicals proper, i.e., to monosemous expressions with variable content like those appearing in Wittgenstein's example above. However, it does not apply to Frege's use of letters, where the idea of the constant meaning of natural language indexicals due to their being just monosemous expressions is rephrased or replaced, rather, in terms of mentioning that the two letters "cross-refer" to one another, even though the same cannot be said about the type/token distinction itself which in the case of Frege's letters—as in the case of letters in general—applies in its most vivid form. Moreover, there is one important distinction that makes these two cases essentially different. The distinction that becomes concealed

[22] *Ibid.*

[23] ...with "meaning" appearing here in the sense of Frege's *Bedeutung*, not *Sinn*.

by Frege's indiscriminate talk about what he calls "indefinite indication" (*Andeutung*) in cases of both indefinite pronouns and letters alike.

As Frege himself explains,

> The distinction between sign and thing signified has not been sharply made, so 'mathematical expression' (*expressio analytica*) has been half taken to mean also that what the expression stands for. Now what does '$x^2 + 3x$' designate? Properly speaking, nothing at all; for the letter 'x' only indicates numbers, and does not designate them. If we replace 'x' by a numeral, we get an expression that designates a number, and so nothing new. Like 'x' itself, '$x^2 + 3x$' only indicates.[24]

What Frege has in mind here by talking about the distinction between sign and thing signified is nothing but what may be called otherwise as *suppositio formalis*. But lacking *suppositio formalis* does not, however, deprive '$x^2 + 3x$' or 'x' of their status of being signs, since what Frege is speaking about is not the distinction between signs and non-signs but the distinction between signs and things signified. And it is quite clear that the complex expression, '$x^2 + 3x$', does not appear in *suppositio formalis*—even though some of its parts do—exactly due to the fact that what does not appear in *suppositio formalis* is one of its component signs, namely the letter 'x'. This, however, points to the fact that the only way in which the letter 'x' in Frege's example above—as well as in any context in which a variable letter appears as part of his constant/variable distinction—may be conceived as a sign is that it appears in one and only one mode that may only be conceivable under these circumstances, namely that it appears in nothing more and nothing less but in *suppositio materialis* (as contrasted with *suppositio formalis* as such).

In its turn, being considered in the light of Peirce's doctrine and, specifically, his type/token distinction, the fact that Frege's inscriptions in the form of uniform letters still have reference—although not in the mode of *suppositio formalis* but *suppositio materialis*—points to the fact that all of Frege's reasoning aimed at the justification of his choice of the constant/variable distinction as the basis of his concept-script and, above all, of his qualification of a variable letter as an expression of generality happens to be based on nothing else but *quaternio terminorum*. For the introduction, by Frege, of the notion of equiformity of letters in expressions such as

(2.21) If a is a man, a is mortal

---

[24] G. Frege, *Was ist eine Funktion*, p. 663.

together with his remark on page 260 of his essay on *Logical Generality* that in (2.21) the equiform letters cross-refer to one another, is aimed explicitly at showing that the letters, since they are equiform, are also co-referential.[25] But here the notion of equiformity implies the notion of co-referentiality to exactly the same extent as the notion of *suppositio materialis* implies, or is able to imply, the notion of *suppositio formalis*. However, since *suppositio materialis* can be shown as only a limiting case of *suppositio formalis*, just in the same way as *Everybody loves himself* can be shown as a limiting case of *Everybody loves somebody*, then it is *suppositio formalis* that can only imply *suppositio materialis*, not *vice versa*. And, specifically, this does not happen in the case of Frege's letters, either. In the first place, this does not happen for the simple reason that, as mere gap signs, they cannot have any *suppositio formalis* whatsoever. This can be shown to be exactly the point that Frege himself makes elsewhere, namely in the passage just cited above:

> The distinction between sign and thing signified has not been sharply made, so 'mathematical expression' (*expressio analytica*) has been half taken to mean also that what the expression stands for. Now what does '$x^2 + 3x$' designate? Properly speaking, nothing at all.[26]

Moreover, his further remark, on page 262, that in sentences like (2.21) he distinguishes the two individual letters equiform with '*a*', from the remaining part, makes it possible to shed light on the exact nature of his *non sequitur*. Admittedly, his concluding remark, namely:

> So I want also to transform the sentence '*If* a *be a man, then is* a *mortal*' into '*If* a *is a man,* a *is mortal*', which in what follows I will call the sentence. In the first sentence I distinguish the two individual letters equiform with '*a*' from the remaining part.[27]

and, specifically, the last sentence of it, seems to have been aimed at providing justification for the constant/variable distinction as the most fundamental distinction of his logical notation. In this remark, the constant/variable distinction reveals itself as the quintessential property of Frege's object-language in which the latter differs from ordinary language such as German, and the introduction of which serves, according to Frege, as an indispensable linguistic tool for rendering generality which otherwise,

[25] ... or, rather, that they are co-referential, since they are equiform.

[26] *Ibid.*

[27] *Ibid.*, p. 262.

without the use of variables, in the eyes of Frege remains imperceptible. This conclusion is based on the notion of the equiformity of letters playing in this distinction the role of variables, and here Frege's statement that the letters cross-refer to one another due to their property of being graphically equiform clearly plays the role of a premise which we cannot but recognise as true and, thus, recognise also as true Frege's conclusion as immediately following from this premise.

Conspicuously enough, the notion of equiformity appears here as a presupposition of Frege's final remark, hence as something already known. However, Frege does not simply reiterate the already-known thesis of his that uniform letters cross-refer to one another, for what he says here is indeed if not entirely new then at least essentially different. For in the preceding remark the point that he is making is that both letters appearing in (2.21) are cross-referring to one another as if it were naturally presupposed by their being equiform with one another quite regardless of the form of the specific letter, '*a*', '*b*', or '*c*', that the letters in question are both made to assume. But in the last remark what appears to be a centrefold issue is not that the two instances of '*a*' are equiform, hence they cross-refer to one another, but—instead—that either of them is equiform with their Type, of which the two letters are only instantiations. Properly speaking, Frege shifts from the notion of mutual equiformity represented diagrammatically as Figure A in (2.22) to an entirely different notion of equiformity with Type, appearing as Figure B in (2.22) below. Quite evidently, Frege thinks it to be a premise (as shown in Figure B) from which it follows, as a conclusion, that the two instances of '*a*' cross-refer to one another (Figure A):

(2.22)

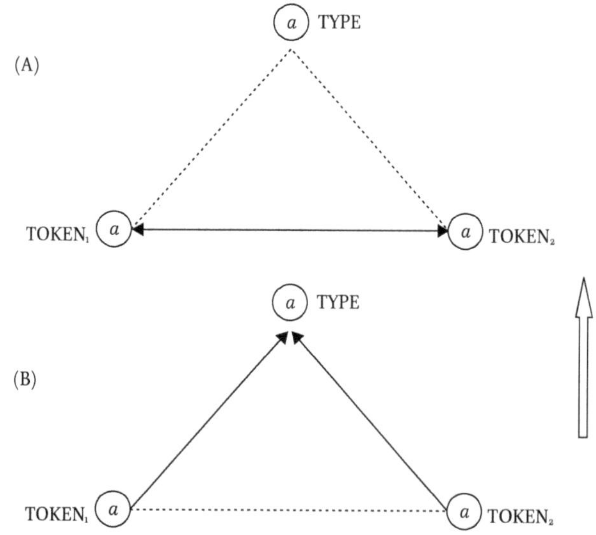

In this way, the covert or, rather, enthymemic character of Frege's reasoning turns out to conceal an evident gap in his train of thought that makes his argument demonstrably false, and which is easily revealed by the Type/Token analysis. For, if we couch Frege's argument in terms of Peirce's Type/Token distinction, we will see that what Frege points at on both occasions while speaking of the two equiform letters being substituted for the indefinitely indicating parts of the sentence in (2.21) are but two different occurrences, hence tokens, of one and the same letter conceived as their Type. But from the fact that he mentions—correctly—in his second remark, namely that the two instances of the letter '*a*' are in fact tokens of the same letter '*a*' as Type and thus either of these letters is equiform with one and the same letter which is their Type (as shown in Figure B above), it does not in the least follow what he conjectures in his previous remark—namely, that the two token-letters cross-refer to one another (as represented by Figure A).

The *quaternio terminorum* that appears in Frege's argument amounts to the fact that he applies two different notions, those of equiformity and cross-reference, as if they were fully interchangeable. Nevertheless, what happens to be true insofar as equiformity is concerned is not at all indicative of the cases involving what Frege terms as co-reference. Indeed, if we look at the diagrams above from these two different angles, it will be seen that in the case of equiformity B entails A, since the token-letter *a* appearing in the antecedent of Frege's conditional in (2.21) is equiform with its Type, and the Type of the token-letter *a* appearing in the antecedent is in its turn equiform with the token-letter *a* appearing in the consequent of (2.21) as the latter is merely a graphical instance of the former, as shown by arrows in Figure B1:

(2.23)

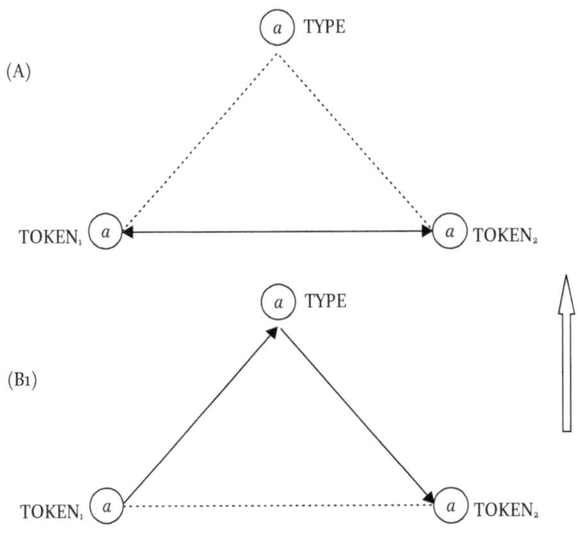

Due to the transitivity of the relation TOKEN-TYPE-TOKEN in (B1) it is plainly equivalent to the respective relation marked by an arrow in (A1):

(2.24)

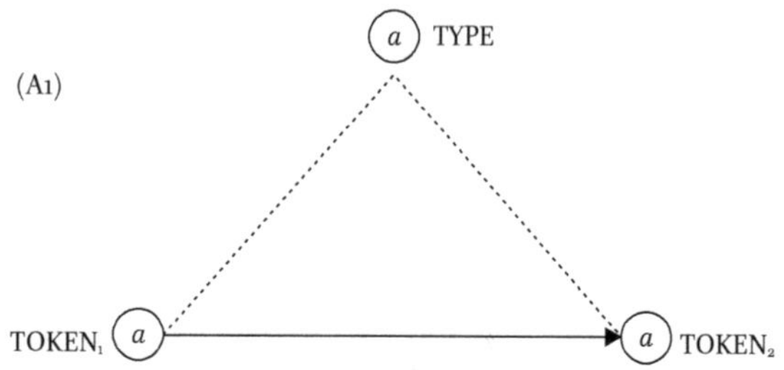

Moreover, since the relation of equiformity is not only transitive but reflexive as well, the same concerns equiformity taken in the opposite direction, as in Figures B2–A2 below:

(4.24)

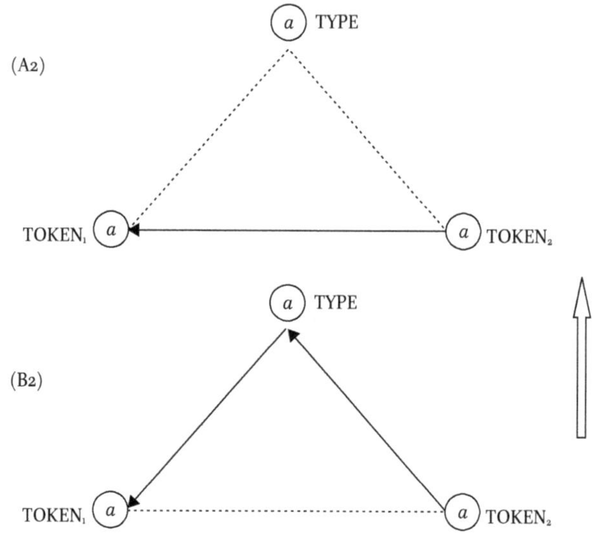

Thus, owing to the reflexivity of the relation in question, we arrive at the equivalent case of mutual equiformity of both tokens, as shown in Fig. 2.23 (A) above.

Both transitivity and reflexivity do not apply, however, insofar as what comes into play are strictly referential properties of the tokens of the letter '*a*' occurring in the antecedent and the consequent of Frege's conditional as well as the letter '*a*' taken as their common type. Notably, this is exactly the case when, to put it in Frege's words, the *expressio analytica* of his object-language is no longer haft taken to mean also that what the expression stands for, and the distinction between sign and thing signified has been sharply made. As the letters in Frege's script do not designate anything, but only, as he puts it, indicate, let us put momentarily aside his notion of indefinite indication[28] to which we will return later, and consider this circumstance (that they do not designate anything) simply as an indication of the fact that they do not appear in *suppositio formalis*; hence, their only mode of designation is that of *suppositio materialis*. Which means that what they only represent or stand for as signs are signs themselves. Notably, it is exactly this mode of *suppositio* that Frege must have at least tacitly assumed while speaking of letters cross-referring to one another since what is clearly at stake when one says that "letters cross-refer to one another" simply cannot be indefinite indication of whatever sort, so to speak, by the very definition itself. Further, the basic fact concerning representation that all signs, hence letters too, perform in the mode of *suppositio materialis* is that they represent themselves or, to put it just differently, that they are signs of themselves. Besides, what is also a basic fact of representation in the mode of *suppositio materialis* is that tokens are representations of their respective types, and not *vice versa*. Thus, any instance of about twenty *the*'s that can ordinarily be found on the written page, for example, appears as a sign of but one word 'the' in the English language as their type. The adverse relation, however, is not possible: the word 'the' in its mode of being as type represents only itself but it does not represent any of its token instances, i.e. it cannot be a sign standing for a token of the same word, for in *suppositio materialis* it may only be a sign of itself, since otherwise the very distinction between Token and Type will be blurred. As a matter of fact, this is implied by the very definition of Token, which Peirce gives in 4.537:

> A single event which happens once and whose identity is limited to that one happening or a Single object or thing which is in some single place at any one instant of time, such as event or thing being significant only as occurring just when and where it does, such as this or that word on a single line of a single page of a single copy of a book, I will venture to call a *Token*.... In order that a Type may be used, it has to be embodied in a *Token which shall be the sign of the Type*.

This far, naturally, also concerns letters insofar as they are considered as signs, too.

[28] ... since, properly speaking, this is a property of the indefinitely indicating parts of the sentence that are only replaced by letters, not of the letters themselves.

Having in view that what now we are after, in the same way as Frege, is nothing but making sharp the distinction between sign and the thing signified so as not to mix our letters of the object-language with what they stand for, and bearing in mind the basic facts regarding signs as Types and signs as Tokens, let us consider the chain of Frege's reasoning strictly from this angle. Since neither transitivity nor reflexivity do not come into play insofar as the *referential* properties of Frege's letters are concerned, the lack of transitivity of reference from the type-letter *a* to the token of the same letter in the consequent will immediately result in the lack of the required transitivity of the referential relations between both Tokens (see Figure B in 2.25). As a corollary, the token letter *a* appearing in the antecedent of Frege's example can no longer be seen as referring to another Token of the same Type appearing in the consequent, as shown in Figure A below:

(2.25)

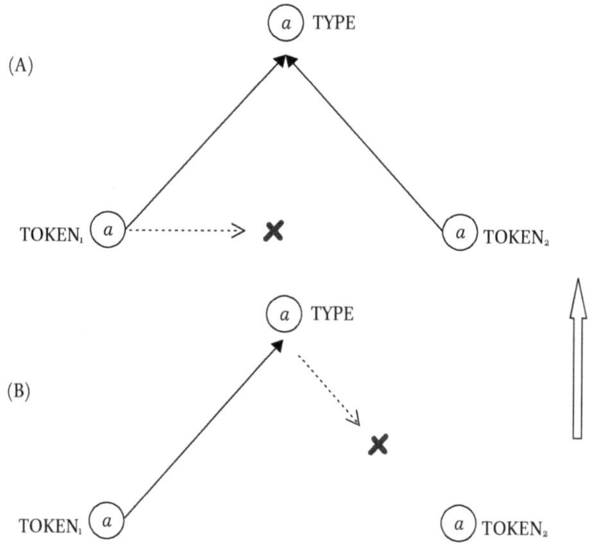

Hence, since neither of the token letters refers to another, they both *do not* cross-refer to one another, either:

(2.26)

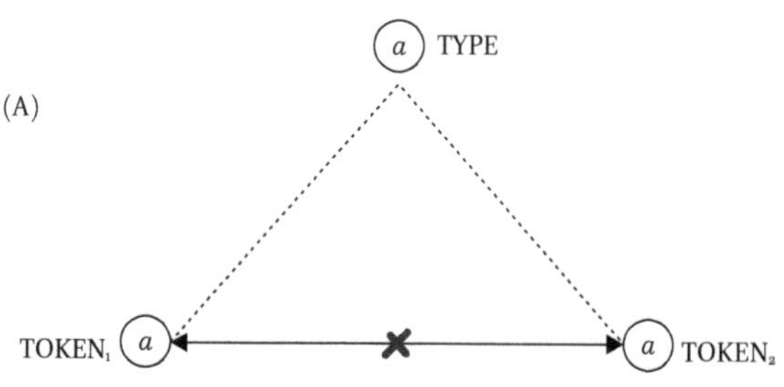

It illustrates any too clearly the nature of Frege's mistake: on his account, Token_1 refers to $\textit{Token}_2$ as if $\textit{Token}_2$ were the *Type* of Token_1, and Token_2 refers to $\textit{Token}_1$ as if Token_1 were the *Type* of Token_2. But here, Type and Token present themselves, so to speak, as two different things: Token is *the sign*, and Type is *the thing signified*. Thus, Frege commits the same mistake on the necessity of avoiding which he himself insists, and on what he insists is the necessity of making sharp the distinction between sign and the thing signified so as not to mix our letters, i.e. variables, of the object language with what they stand for. But, in Frege's framework, this mistake is unavoidable. For what stands for generality is clearly Type only. However, what Frege tacitly assumes—and in his train of thought this assumption is quite indispensable since it is the most basic assumption of Frege's philosophy upon which the very idea of his *Begriffsschrift* comes ultimately to depend—is something that *cannot* be substantiated: namely, that *being* an expression of generality is wholly interchangeable with *standing for* such an expression. For, generally speaking, this amounts to assuming no more and no less than that the word 'pudding' is the very same thing as pudding itself.

3.3. *The Generality of Reference and the Reference of Generality*

Why, then, is it so that despite the outwardly obvious fact that both occurrences of a proper name such as 'Napoleon' in

(3.31) If *Napoleon* is a man, *Napoleon* is mortal

do however seem to cross-refer to one another, the same does not happen neither in

(3.32) If *a* is a man, *a* is mortal

nor in

(3.33) If *something* is a man, *something* is mortal

even though in (3.32)–(3.33) the respective *indefinitely indicating* expressions do happen to be equiform? Moreover, what can be easily seen if we compare (3.33) with

(3.34) If *something* is a man, *it* is mortal

is that it is exactly the equiformity of the both occurrences of 'something' which happens to be standing in the way of their being able to "cross-refer" to one another. Quite notably, what the examples like (3.33)–(3.34) readily exemplify is that *indefinitely indicating* expressions—as opposed to proper names like 'Napoleon' that enter their structural positions by substitution—may either be equiform or they may cross-refer to one another, but they cannot *both* be equiform and cross-refer to one another. And, in fact, this goes as a clear indication that the regularity that we may observe in the case of indefinitely indicating expressions in *suppositio materialis* happens to hold in *suppositio formalis* as well. What is more, this gets further independent support on a thorough observation of quite various syntactic phenomena extensively discussed a few decades ago within the framework of what is now known as early generative grammar. For, in fact, the case exemplified by (3.33)–(3.34) happens to be not quite exceptional as it makes only a minor instance of the notoriously known issue arising in the framework of transformational grammar, namely that of whether transformations can change meaning.

Representative examples of such meaning-changing transformations, as pointed out by B. Partee and J. Hintikka,[29] are the following:

(i) pronominalization:
   a. Some boy believes that Mary loves him.
   b. Some boy believes that Mary loves some boy.

(ii) reflexivization:
   a. Every man voted for himself.
   b. Every man voted for every man.

(iii) equi-NP deletion:
   a. Every contestant expected to come in first.
   b. Every contestant expected every contestant to come in first.

(iv) relative clause formation:
   a. Every Democrat who voted for a Republican was sorry.
   b. Every Democrat voted for a Republican, and every Democrat was sorry.

(v) conjunction reduction:
   a. Few students are both popular and likely to succeed.

[29] J. Hintikka, *Quantifiers in Logic and Quantifiers in Natural Languages*, pp. 211–12; B. Partee, *On the Requirement that Transformations Preserve Meaning*, pp. 9–10.

b. Few students are popular and few students are likely to succeed.

In the examples above, (a) and (b) sentences are clearly not synonymous. Such discrepancies required explanations, and it is also evident that explanations could not have been readily provided, as on Chomsky's early proposal,[30] by mere reference to the extraneous factors, manifested by the order of quantifiers in surface structures, since we can hardly speak of any order of quantifiers in the surface structure of (1a)–(5a) at all while in (1b)–(5b) quantifiers are mutually interchangeable, as part of identical NPs, without any semantic consequences whatsoever. Hence, what examples of this kind might most probably mean is that either transformations (i)–(v) fail to preserve meaning or else (a) and (b) sentences in such examples have different underlying structures so that they are not transformationally related—of which neither could be accepted as a plausible explanation since both went contrary to the basic formal assumptions of the theory.

However, a little bit closer examination of the examples of such meaning-changing transformations will readily show that what was then known as the problem of quantifiers in generative grammar happens to be touching upon the issue that is relevant to the foundations of logic as the examples of this kind are very likely to pose serious problems on Frege's approach, too. Because what they make most vividly manifest is not only the failure of the early generative approach to account for the most regular transformational relations between sentences, but also the failure of Frege's notion of equiformity as a necessary pre-requisite of referential identity, since what all of the examples above also show is nothing but the fact that whenever indefinitely indicating expressions are equiform, they obligatorily *do not* cross-refer to one another. Thus, this phenomenon happens on a quite regular, lawlike basis and not by way of an exception to a general rule. For, notably, the very fact that in (iii)–(v), as opposed to (i)–(ii), the required coreference is attained not by merely dispensing with the equiformity of the indefinitely indication expressions but also by making obligatory the deletion of the equiform indefinitely indicating expression itself[31] is not in the least exceptional, but makes the fact that

[30] N. Chomsky, *Aspects of the Theory of Syntax*, p. 224.

[31] However, the use of the term 'deletion' might itself make the case of the use of the wrong idiom since what the observed phenomena may also suggest is that the rule is inapplicable at the level of deep structure as well and thus the items in question are not in a

indefinitely indicating expressions, if they are equiform, then also mutually exclusive, a general rule.

The inadequacy of the very basics of Frege's doctrine becomes, however, even more evident since this raises a host of yet further questions, in providing any coherent answers to which Frege's formal notion of coreference as equiformity proves to be entirely and most utterly recalcitrant. Thus, if we were to treat the letters in (3.33) as fully substitutable with logical proper names and therefore to treat them *as if* they were proper names themselves, should we then not be in a position to claim that since in all cases of identities involving singular terms like 'The evening star is the same as the morning star' both terms like 'evening star' and 'morning star' refer to a single object, then they also cross-refer, hence, are equiform? But shouldn't it then follow that in a logically well-behaved, or in a purely logical notation such cases should be symbolized as $a = a$, whereas their symbolization in the form of $a = b$ should be regarded as only representing the effect of the way that natural language, due to its logical defectiveness, happens to obscure "pure thought"? In a truly logical notation, then, all symbolizations of the form $a = b$ should be simply banned as an unwanted remnant of a logically corrupt language. For, in the language of "pure thought," they should be only remedied, but not analysed since otherwise it would be an analysis of something that cannot truly count as logical matter in much the same way as the subject/predicate distinction itself.

And if *Beggriffsschrift* as our object-language is so construed as to be able to make the true structure of thought perspicuous, since otherwise it is distorted by the "vagaries" of the linguistic structure of ordinary language, isn't it clear that this can only be done if and only if it renders the structure of thought in a one-to-one manner? For, if in

(3.32) If a is *a* man, then *a* is mortal,

where the *a*'s are but "constant" letters representing the name 'Napoleon', we can see nothing standing in the way of the two instances of the constant letter *a* to cross-refer to one another just the way that both occurrences of 'Napoleon' do, how can the matter be different in

position of leaving something like unobservable 'traces' at the observable level of surface structure as the fact of their "deletion" is only putative. For more on putative "deletion," see Section 4.5.

(3.35) If x is a man, then x is mortal

after we have substituted the "variable" letter x for the "constant" letter a? If (3.32) is just an instance of a general law as represented in (3.35), and both cases, by definition, represent a general thought appearing in one of its particular cases and the general thought itself, accordingly, in a *one-to-one* manner, why shouldn't it follow with all logical necessity that the x's appearing in the representation of the general thought also cross-refer to one another?

Since, however, what we expect does not really follow from these premises, even though the line of reasoning appears to be impeccable, from all this follows a tiny but, indeed, extremely important consequence. On a closer examination it becomes evident that the source of error lies not in a *logical* fallacy, but in a fallacy which is purely *material*, arising from the falsehood of one of the premises. Namely, from the assumption by Frege of the one-to-one manner in which his symbolic representation allegedly reflects thought and, ultimately, from Frege's complete reliance on the wrong idiom of "*reflection* of thought" on which he rests his negative consideration that language obscures thought by providing its reflection that is distorted and impure. Because, in the first place, it is just this premise concerning the alleged capability of rendering, by means of his *Begriffsschrift*, thought, or else reflecting it, in a one-to-one manner that contradicts what Frege himself claims elsewhere, namely that

> ...we must not fail to recognize that the same sense, the same thought, may be variously expressed;...If all transformation of the expression were forbidden on the plea that this would alter the content as well, logic would simply be crippled; for the task of logic can hardly be performed without trying to recognize the thought in its manifold guises. Moreover, all definitions would then have to be rejected as false.[32]

It will be seen, however, that to recognize that "the task of logic can hardly be performed without trying to recognize the thought in its manifold guises" would be very hard indeed unless we also recognize that rendering thought in a one-to-one manner is equally unattainable. Moreover, unless what we also recognize is that it is unattainable, equally, both in natural language, where it is deemed to be happening due to its alleged capability of obscuring thought, and in Frege's *object-language itself*. Notably, the latter is clearly taking place quite despite the fact that "rendering thought in

[32] G. Frege, *Über Begriff und Gegenstand*, p. 196.

a clear and precise way" is nothing but exactly the purpose of his *Begriffsschrift*. And this may mean one thing only, namely, that thought cannot be "reflected" by language cannot be explained away simply by referring to its alleged logical defects, not even in the case of Frege's logical notation as such. For what we are easily tempted to take for a deficiency in fact turns up to be a regularity that is not properly understood and, thus, only *conceived*, falsely, as its deficiency. The deficiency in question does not therefore belong to language as such; hence, what it properly belongs to is only the way that we happen to understand language. Thus, there is no way, anymore, of speaking about language being logically defective; rather, what really happens to be logically defective is our own apprehension of it.

Notably, the fact that what Frege terms as "rendering thought in a clear and precise way" is impossible, not even by way of its "reflection" by means of Frege's *Begriffsschrift*, has far-reaching and, indeed, fundamental implications. For if it cannot be done in principle, then it can be for one reason only—namely, for the reason that thought and language may not or, rather, cannot be such separate things or entities so that one could be the sign, "distorted" or perfect, of another. Hence, whatever they can be, they can only be one and the same entity. Or, to say it differently, whatever this entity may be, it can only be an entity of such kind in which language and thought can only be seen as entirely inseparable—no question of proper or else improper reflection can therefore ever arise. For there can be no question of reflection between that which simply *cannot* be separated. Then, as a matter of fact, the only aspect in which proper or improper reflection can only be conceived is our own reflection, proper or improper, *upon* this one single entity that is both thought and language in one. But the conclusion that we thus arrive at happens to be roughly equivalent to one of Peirce's four well-known denials summing up his fundamental approach:—"We have no power of thinking without signs" (5.266). In fact, one of the basic principles of Peirce's philosophy, if not the most basic one—namely that thought is inconceivable without signs—appears to be correct, and it appears to be correct on a very specific, ontological reading: that thought *does not exist* without signs. And if without signs thought does not exist, then it also cannot be something that exists *prior* to signs. Hence, there simply is nothing in thought for signs to "reflect" in the sense that the idiom of "reflection" might suggest, and the whole issue of signs reflecting thought, properly or improperly, proves to be nothing but a mere misnomer. But then there can be no such thing as the logical structure of the world, such that it could be "reflected" by the logical structure of our language, for all that logical structure can ever be is the structure of the sign, and nothing but the sign, itself.[33]

3.4. *Peirce's Real Dyad and Causality*

In his introduction to Wittgenstein's *Tractatus* dating back as far as 1922 Bertrand Russell wrote:

> There are various problems as regards language. First, there is the problem what actually occurs in our minds when we use language with the intention of meaning something by it; this problem belongs to psychology. Secondly, there is the problem as to what is the relation between thoughts, words or sentences, and that which they refer or mean; this problem belongs to epistemology. Thirdly, there is the problem of using sentences so as to convey truth rather than falsehood; this belongs to special sciences dealing with the subject-matter of the sentences in question. Fourthly, there is the question: what relation must one fact (such as a sentence) have to another in order to be *capable* of being a symbol for that other? This last is a logical question.[34]

A little bit earlier such question would have been categorised as a one of philosophical grammar, or rather—as a question having direct relevance to Peirce's doctrine of *logic as semiotic*. Here, Russell explicitly places the relation between a fact (such that it may also turn up to be a sentence) to another within the context of the general theory of signs. And what Russell is genuinely concerned here with is the question about the material necessary conditions for one fact such as a sentence to appear as a symbol for another fact. Or, if we want to dress it in semiotic terms—about the material necessary conditions for one entity (object or event) of being a sign for that other.

In the general framework of Peirce's theory of signs such entities, namely facts appearing as signs of still other facts, fall under one particular category, where the sign, as it is in itself, is categorised as a *sinsign*[35] or, according to its mode of presentation, an *Actisign*,[36] or as Peirce defines it elsewhere, a *Token*.[37] Taken more generally, this latter characterises only

---

33 For all that logical structure can—and indeed, by its very definition, should—ever be is the structure of the λογοσ, i.e. logical discourse. "However, what symbolic logic furnishes is anything but a logic, i.e. a contemplation of the λογοσ" (M. Heidegger, *Die Frage nach dem Ding*, p. 122).

34 L. Wittgenstein, *Tractatus Logico-Philosophicus*, Preface by Bertrand Russell, pp. 7 ff.

35 *CP* 8.334.

36 *CP* 8.347.

37 *CP* 8.364.

one of the three possible modes of being of the sign, namely those of *Mark* (or *Tone*), *Token*, and *Type*. In the case of Russell's sentences, Peirce's *type-token-tone* distinction would then involve a distinction between the sentence as an abstract type, the linguistic event, and the actual physical inscription, the mark on the paper, the physical sound waves etc.[38]

Notice, however, that this distinction is not intended to speak to the relation between the sign and its object (that other fact), but primarily to the modes of being of the sign *qua* object (or fact itself). In order to consider the sign in its most fundamental sign-relation, and thus in the mode of being of the sign *qua* sign (and not as merely *object*), we need to consider it as appearing, in Peirce's terminology, as an *Index*. The latter, for Peirce, is either 'really connected' with its object, or else it indicates its object 'independent[ly] of the mind using the sign'.[39] As such, it appears as part of the trichotomy of icon, index and symbol, which Peirce defines in 2.247–9 as follows:

> 2.247. An *Icon* is a sign which refers to the Object that it denotes merely by virtue of characters of its own, and which it possesses, just the same, whether any such Object actually exists or not. It is true that unless there really is such an Object, the Icon does not act as a sign: but this has nothing to do with its character as a sign. Anything whatever, be it quality, existent individual, or law, is an Icon of anything, in so far as it is like that thing and used as a sign of it. 2.248. An *Index* is a sign which refers to the Object that it denotes by virtue of being really affected by that Object. It cannot, therefore, be a Qualisign, because qualities are whatever they are independently of anything else. In so far as the Index is affected by the Object, it necessarily has some quality in common with the Object, and it is with respect to these that it refers to the Object. It does, therefore, involve a sort of Icon, although an Icon of a particular kind; and it is not the mere resemblance of its Object, even in these respects which makes it a sign, but it is the actual modification of it by the Object. 2.249. A symbol is a sign which refers to the Object that it denotes by virtue of a law, usually an association of general ideas, which operates to cause the Symbol to be interpreted as referring to that Object. It is thus itself a general type or law, that is, a Legisign. As such it acts through a replica. Not only is it general itself, but the Object to which it refers is of general nature. Now that which is general has its being in the instances which it will determine. There must, therefore, be existent instances of what the symbol denotes, although we must here understand by "existent", existent in the possibly imaginary universe to which the Symbol refers. The Symbol will indirectly, through the association or other law, be affected by

[38] Cf. C. Hutton, *Abstraction and Instance*, p. 20.
[39] *CP* 3.361.

> those instances; and thus the Symbol will involve a sort of Index, although an Index of a peculiar kind. It will not, however, be by any means true that the slight effect upon the Symbol of those instances accounts for the significant character of the Symbol.

Consideration of the definitions that Peirce lays down in 2.247–9 exposes the very point that makes Russell's logical question:

> What relation must one fact (such as a sentence) have to another in order to be *capable* of being a symbol for that other?[40]

both challenging and complex. In fact, Russell's logical problem, as it appears from the perspective of Peirce's conception of *logic as semiotic*, happens to be deeply rooted in a clear tension that arises, on one reading of Peirce's writings, between the "natural" and the "conventionalist" interpretations of the nature of the relation between sign and object. The intricate character of this tension can be best shown by the following citation from Hutton:

> Peirce shows a fundamental mistrust of all theories of representation that do not rely on some intrinsic or natural or non-conventional relation between the sign and its object. This is the underlying rationale for the statement that it is only through natural (i.e. not purely conventional) relations that ideas are communicated, for the inclusion of an iconic element in indexicality (2.248) and for the inclusion of an element of an indexicality in the definition of the symbol. The icon 'represents' in virtue of showing or resembling its object; the index operates 'where one thing becomes the sign of another through being connected with it by some natural law or tendency' (Ayer 1968: 152). Smoke is not arbitrarily related to fire, it is a 'natural' sign of it.
>
> Against this we must set Peirce's recognition that a strong element of conventionality is present in all representation. This can be seen clearly from Peirce's discussion of the conventionality and intentional nature of icons, and from his distinction between icon and hypoicon. The drawing of this distinction is predicated on the idea that resemblance can only be conceived of from the point of view of an observer or interpreter who perceives it. These two strands—the 'naturalistic' and the 'conventionalist' views of representation—combine in the account of algebraic formulas as icons. This notion of iconicity is reminiscent of Wittgenstein's 'picture theory' of meaning, expounded in the *Tractatus* (1922).
>
> This duality is reflected in a further ambiguity in Peirce's theory of signs. If we consider the definition of the sign as 'something which stands to somebody for something' (2.228), the question arises: does every sign need an

[40] ...which is fundamentally Kant's own question about the conditions of reference to objectivity.

> interpreter? And if so, are we talking of type or token, of generalized or actual reference? The importance of this can be seen when we consider the issue of the natural vs. conventional nature of the sign. As Ayer (1968: 157) points out, Peirce often neglects 'the fact that iconic and indexical signs stand as much as any others in need of interpretation'. If we mean by interpretation, the act of an individual on a particular occasion, we have to be sure that every individual would interpret a particular sign as being in a certain iconic relation to a certain object. If this condition cannot be met, then we cannot speak of a type-token relation independently of interpretation in the case of icons and indices, any more than with conventional symbols (which by their very conventional nature must depend on an intermediate interpreter.[41]

As Hutton further points out:

> Peirce is faced with a dilemma: to assert the conventional nature of icons and indices is to deny his semiotics a base in 'natural' representation. The consequence of this will be that some account must be given of the relation of sign to interpreter. For one of the ways of conceiving of naturally iconic or naturally indexical signs is as in principle independent of the interpretive act of any individual. The individual has no choice but to perceive the resemblance or the relation. But it is in a 'conventional' sign-system that the distinction between type and token is hardest to draw. If the token were 'naturally' iconic of the type (or, more logically, of other tokens), then the notion of interpreter could be either dispensed with, or—something which comes to mean the same—assumed to be context-independent. But if the relations within the sign-system are conventional, then some act of perception or application must be allowed for. To say that the token is related by convention to the type, or to other tokens, is to transfer the problem from the signs themselves to the interpretative act(s).[42]

There is, however, a way of invoking the notion of interpreter without the need to transfer the problem from the signs themselves to the interpretative act(s). For by his very philosophy of categories, where Secondness is defined against Firstness, Peirce in fact provides grounds for giving such an account of the relation of sign to interpreter quite *without* denying his semiotics a base in 'natural' representation:

> We need not, and must not, banish the idea of the first from the second. On the contrary, the second is precisely that which cannot be without the first [...]. A thing cannot be other, negative, or independent without a first to or of which it shall be other.[43]

---

[41] C. Hutton, *Abstraction and Instance*, p. 28.
[42] *Ibid.*, pp. 28–29.
[43] *CP* 1.358.

The idea of Secondness ('existence') as defined against Firstness (simply 'being' as opposed to 'existence') is exactly that of a sign *qua* sign, when a sign is conceived not as an object, but through its relation to another object, as a sign, i.e. as 'something which stands to somebody for something' (2.228).

It will be helpful to consider this basic sign relation, that of 'standing [to somebody] for something' in the light of Peirce's further philosophical distinction that he draws between a real Second and a 'degenerate' Second. In both cases, the base in 'natural' representation is secured by the necessity for the very existence or, rather, 'being' of that other object due or owing to which the very mode of the existence of sign *qua* sign is ever possible. And the distinction between a real Second and a 'degenerate' Second appears to secure that natural, objective 'base' which finds its locus in Peirce's philosophical category of Secondness, with respect to which the role of the interpreter turns up itself to be only derivative, in as much as it appears to depend upon a further subdivision—within the purely existential, hence 'material' Secondness—which Peirce coins in terms of the opposition between the 'real' Secondness and the 'degenerate' one. This differs, of course, from Frege's view on being as something that enters into causal chains, from which his logical notion of existence (i.e. of the concept that is *not empty*) follows, in effect, as a corollary. The effect of duality, or tension arises, therefore, only when Peirce's sign relations are being thrust upon Frege's, but not Peirce's own, philosophical understanding of existentiality.

In Peirce's distinction between a real Second and a 'degenerate' one that he sets forth in (1.365), a degenerate Second is a 'relation of reason', a 'mere aggregate of two facts'. For example, Rumford and Franklin 'resembled each other by virtue of being both Americans, but either would have been just as much an American if the other had never lived'. A 'real' Second is found in some indivisible dyad: 'the fact that Cain killed Abel cannot be stated as a mere aggregate of two facts'. As is easy to notice, the basis of this distinction formed by the respective distinction between 'degenerate' *vs*. 'real' dyad lies in the opposition between a dyad having a character of being a mere aggregate, on the one hand, and a respective dyad devoid of its character of being a mere aggregate, on the other. And, as can be clearly seen from Peirce's examples, the matter of the two facts not being a mere aggregate or being such is itself the result of whether they enter into a causal chain or not.

Philosophical category:

'Degenerate' Second	'Real' Second

Type of respective Sign-Object relation:

'Degenerate' dyad	'Real' dyad

Sign-Object relation as a whole characterised as:

Being a mere aggregate	*Not being* a mere aggregate

Sign-Object relation characterised with respect to the presence of a causal chain (CC):

Sign-Object relation ***NOT formed*** by a CC	Sign-Object relation *formed* by a CC

The characterisation of the degenerate Second as a 'relation of reason' then goes in fact as a corollary of, and is only contingent on, the inherently natural or else purely physical basis of Peirce's distinction since, as Peirce himself explicitly remarks: "Any two objects in nature resemble each other, and indeed in themselves just as much as any other two; it is only with reference to our senses and needs that one resemblance counts for more than another."[44] Thus, the material necessary condition for the sign being a sign is not, therefore, in its connection with the referent of the sign through the mediating mind as appearing in (4.41):

(4.41)

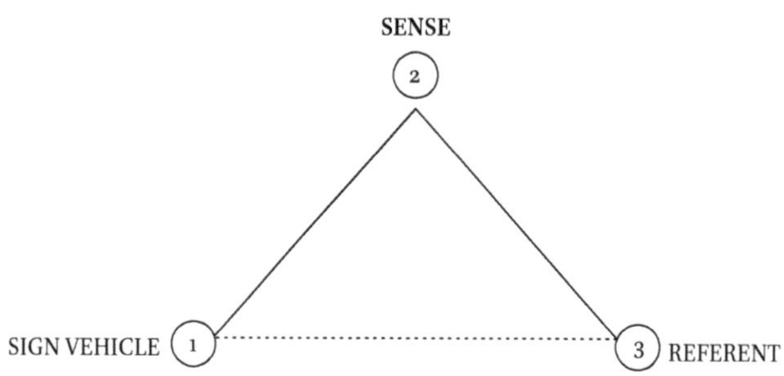

in the semiotic tradition following Gomperz[45] and Odgen & Richards,[46] which, however, rests on a very specific reading of Peirce's texts or, rather, interpretation of his terms, according to which the use of the terms 'genuine triad' and 'genuine sign' for the characterisation of the so-called semi-

[44] *Ibid.*, 1.365.
[45] H. Gomperz, *Weltanschauungslehre*.
[46] C. Ogden, and I. Richards, *The Meaning of Meaning*.

otic triangle in (4.41) would suggest the presence of the path of mediation as showing itself as the basic constitutive feature of the sign *per se*. It is exactly the specific context of such presuppositions widely reiterated further in the literature[47] that gives certain authority to conclusions, such as Silverman's, that

> The signifying value of the weathervane resides not in its physical relationship to wind, but in the concepts 'wind' and 'direction' which it enables the observer to link up.[48]

In Peirce's terms the signifying value, however, does not seem to reside in the concepts 'wind' and 'direction' which it enables the observer to link up. In the first place, the weathervane is a sign to that very extent in which it represents Secondness and, as such, is conceived as a thing that "cannot be other, negative, or independent without a first to or of which it shall be other." In other words, the weathervane is something that is not wind itself. It is something that is other than the wind, and only thus it can come into being as a sign of the wind. For the weathervane, as such, is Firstness. So is the wind. And it is only in its binary relation to the latter that the weathervane also acquires the status of Secondness and thus as something that exists due to the fact that its existence is not that of the wind. For the simple reason that otherwise it would not be the existence of the weathervane but that of the wind itself—the existence of anything but the wind itself would then be simply forbidden. Therefore, whatever similarity there can be, it can arise only on the basis of difference. The notion of difference, or otherness, is thus the basic element in Piece's thinking about signs in much the same way as it was in the work of Saussure, the distinction being in that Peirce considered otherness as the basic characteristic of the relation between signs and non-signs, whereas Saussure, of the relation between signs themselves.

Thus, in Peirce's thinking, otherness, as well as similarity (as that against which otherness can only show itself as such),[49] appear both to be equally basic constituting features of any Second. "Any two objects in nature resemble each other, and indeed in themselves just as much

[47] Cf., e.g., J. Lyons, *Semantics*, p. 96; H. Lieb, "Das ‚semiotische Dreieck' bei Ogden und Richards: eine Neuformuliering des Zeichenmodells von Aristoteles". In H. Geckeler, et al., eds., *Logos Semanticos*, pp. 137–56.

[48] K. Silverman, *The Subject of Semiotics*, p. 19.

[49] Cf.: "The forms of the word *similarity* and *dissimilarity* suggest that one is the negative of the other, which is absurd, since everything is both similar and dissimilar to everything else" (*CP* 1.567).

as any other two."[50] In other words—any object is already a sign of any other with regard to which it emerges as the Second in virtue of its otherness of, as well as its resemblance to, that other object in one respect or another. But it is only due to its prior existence in nature, such as the resemblance of the direction of the weathervane with that of the wind with respect to which the weathervane acts as an icon, that one of quite a multitude of resemblances counts for more than another, and it is only by way of complementation, 'with reference to our senses and needs', as Peirce notes, that it does so. And the signifying value of the weathervane is thus self-subsistent in any of the instances of its emerging as the Second, quite regardless of the external observer who would, depending on reference to his senses and needs, either reckon it as the Second or leave it entirely unnoticed.

Returning to the semiotic triangle, we may thus see that the graph in (4.41) misses this important generalisation of Peirce. For, in the vein of his philosophy, the relevant sign-constituting dyad emerges as the one that is formed by the mind-independent elements connected by the bottom line in (4.42) below, with respect to which the role of the observer/interpreter appears to be entirely complementary:

(4.42)

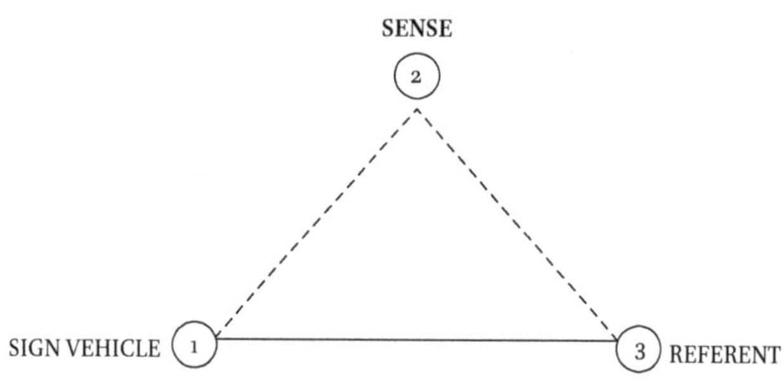

But, being complementary to the basic link between the sign and the object, the path of mediation from (1) to (3) appears to be thus crucially dependent on this basic link, and not *vice versa*. For its very status as an instance of the philosophical Secondness is that which already qualifies it as a self-sustained sign-relation and, thus, quite independent of any mediation on the part of the observer. And this, in fact, is what Peirce stresses himself in his characterisation of an index which, according to

[50] *CP* 1.365.

Peirce, is either 'really connected' with its object, or else it indicates its object 'independent[ly] of the mind using the sign'.[51] In Peirce's terms, saying that this dyad, the sign-object pair, has Secondness as its mode of being amounts to saying that the two together constitute a brute fact, whose reality is entirely independent of whether or not anyone uses it in a sign. (5.73)[52] That is, for Peirce, to say that the [indexical] sign indicates its object independently of the mind using the sign is no more than just another merely equivalent way of stating that

> ... the relation of the [indexical] sign to its objects consists ... in some existential relation to that object ...[53]

Viewed against this background, the sign-object dyad will always be seen as one that is formed by this existential relation, split into its two kinds, or varieties—a) an existential relation expressing causality, as in the case of a real, or genuine dyad, and b) an existential relation in which the causal character of the sign-object pair is denied, as in the case of a degenerate dyad. The denial of causality in the latter case will then have nothing to do with whatever denial of the existential character of the relationship within the sign-object pair, provided that an index, which in this case does not happen to be causally connected with its object, still indicates its object 'independent[ly] of the mind using the sign'. The denial of causality, therefore, will not be the denial of existentiality, unless the independence of the sign user is also denied. For the very denial, in the latter case, of the sign's independence from the mind using the sign will then amount, roughly, to saying that the basic sign relation—hence, the sign as such—does not exist. Note that from this latter circumstance it clearly follows that the denial of causality will also turn out to be the denial of existentiality in one case only—when the denial of causality happens to be applied simultaneously with the denial of the independence of the basic sign relation from the sign user. In other words, if it goes along with the statement that this basic sign relation happens to be *dependent* on the mind using the sign.

In view of this latter consideration, the wording of Peirce's distinction between indices formed by a real dyad, on the one hand, and by a degenerate one, on the other, would be somewhat misleading since in fact this distinction happens to be based not on the opposition between a *real*

[51] *CP* 3.361.
[52] Cf. D. Pharies, *Charles S. Peirce and the Linguistic Sign*, p. 39.
[53] *CP* 2.243.

dyad and a degenerate dyad as a *non-real* one but, rather, on a more down to the matter of fact distinction between the *reality of a genuine* dyad as opposed to the same kind of *reality of a degenerate* one. In either of these cases, the relation of the indexical sign to its objects consists in some existential relation to that object. Hence, in both of these cases, indices are to the same extent existent, or real. In this specific sense, all kinds of indices *are* real. And, as such, in both of these cases they do indicate their objects 'independently of the mind using the sign'.

Having dispelled, and thus dispensed with, the false sense of a degenerate dyad, in which it would appear as if it were something '*non*-real', as contrasted with—and in this way mistakenly opposed to—the brute reality of a 'real', or genuine dyad, we need also to consider the underlying sense of Peirce's use of the terms "genuine" and "degenerate" as it stands with regard to the meaning of causality. As it was already noted earlier, Peirce's understanding of *existence* in terms of Secondness, as opposed to Firstness as *being*, happens not to be based, as is in the case of Frege, on whatever involvement of causality. Quite on the contrary, for Peirce 'existence' is being 'attained' quite regardless of whatever else and, thus, *independently of*, and in fact *prior to* causality as such. Just a quick look at various terms used by Peirce in his depiction of a sign-object relation shows that the basic characterisation of the relation in question happens to be the one in which the sign and its object are in the state of being 'a mere aggregate' which he uses in the definition of a degenerate dyad. Or, rather, the idea of 'being an aggregate', to which 'mere' as an attribute is adduced so as to mark its most simple, basic form, whereas the negation of 'mere' on the right side of the table below points to the composite, and thus derivative character of the sign-object relation within the dyad that is 'real' or genuine':

Philosophical category:

'Degenerate' Second	'Genuine' Second

Type of respective Sign-Object relation:

'Degenerate' dyad	'Genuine' dyad

Sign-Object relation characterised with respect to the presence of a causal chain (CC):

Sign-Object relation *not* formed by a CC	Sign-Object relation *formed* by a CC

Sign-Object relation as a whole characterised as:

Being a ***MERE*** aggregate	***NOT* being a *MERE* aggregate**

The sign-object relation formed by a causal chain, appearing in the case of a genuine dyad, presents itself as the result of a brute negation of the fact that the sign and its object are in the state of being a 'mere aggregate'. And, conversely, the sign-object relation in its most primitive, basic state of being a 'mere aggregate' may also be seen as a sort of existence with the status of unattained causality, the attainment of which depends crucially on the negation, or denial of the fact that the sign and its object are in the state of being a 'mere aggregate'. And thus, by this very fact, it depends on stating that there is *something more* to their state of being an aggregate—which prevents, or rather adds something to, its *pure* characterisation as a state of being a '*mere* aggregate', or of being 'merely an aggregate'.

3.5. *A Dual Perspective on Causality and Mind-Independence*

In this pure, primordial state of being as a sign-dyad, its existence is therefore determined prior to any interpretation, which manifests itself in Peirce's dictum that the sign indicates its object 'independent[ly] of the mind using the sign'. The objective, mind-independent character of the sign-dyad follows directly from its status as an instance of the philosophical Second. With respect to this instance, it then holds that:

> To say that something has a mode of being which lies not in itself but in its being over against a second thing, is to say that that mode of being is the existence which belongs to fact.[54]

The mind-independent character of the sign, which boils down to the being of the sign (vehicle) as something which exists in its being over against a second thing (sign's referent), appears therefore as a material necessary condition of the sign *qua* sign in either of the two varieties of Secondness. Namely, in the case when the sign-relation is made up by a genuine dyad as an instance of the philosophical Second which—to use Gallie's wording—is the conception *of reaction with* something else as well as, equally so, in the case of a degenerate dyad as an instance of the philosophical Second being versed as the conception *of* (pure) *being relative to* something else.

The distinction between these two modes of Secondness as well as their mutual interrelationship as shown in the following table, calls, however, for closer examination and much more thorough consideration:

[54] *CP* 1.432.

Existence	Being a (mere) aggregate	Degenerate sign
Causality	Being more than a (mere) aggregate	Genuine sign

Considering causation as the denial of being 'merely an aggregate' and discerning the ways of attaining the latter (the denial of being 'merely an aggregate') both in the case of 'reaction with' and in the case of 'being relative to'—which go parallel to the twofold characterisation of an index which, for Peirce, is either 'really connected' with its object, or else it indicates its object 'independent[ly] of the mind using the sign'[55]—we find the relevant case in the second row representing the case of a conventional sign in the table below:

Causality/Primary/ Non-Derivative (Real, as *Not* Degenerate)	Being more than a (mere) aggregate	Genuine (real) dyad *qua* natural Index, as a result of negating the state of being a mere aggregate *due to the presence of causal connection*
Causality/Secondary/ Derivative (Real, as *Non*-Degenerate)	Being more than a (mere) aggregate	Degenerate dyad *qua* conventional Index, as a result of negating the state of being a mere aggregate *due to indicating its object independently of the mind using the sign*

In this latter case the relation of the [indexical] sign to its object consists in some matter-of-fact existential relation to that object. In a way, this is an equivalent way of saying that the indexical sign in question indicates its object independently of the mind using the sign. In this case the lack of any straight causal or natural connection is conditioned by its transcendental character that does not show itself in the sphere of experience that may therefore point to the presence of some functional equivalent of such connection due to which the indexical sign in question is capable of indicating its object, in which the mind using the sign cannot be engaged. The purely logical way to shed light on the issue would therefore amount to accepting or else entertaining the assumption that this functional equivalent is based on a a possibility of diverse ways of refuting the state of being a mere aggregate, and thus on the presence of some (non-classical)

---

[55] *CP* 3.361.

sort of dual negation, leading to the effect that is similar to the one that we find in the case of a genuine dyad. Graphically, this difference may be represented via the opposite direction of negation applied to the state of being a mere aggregate as its starting point in Fig. III:

Fig. III.

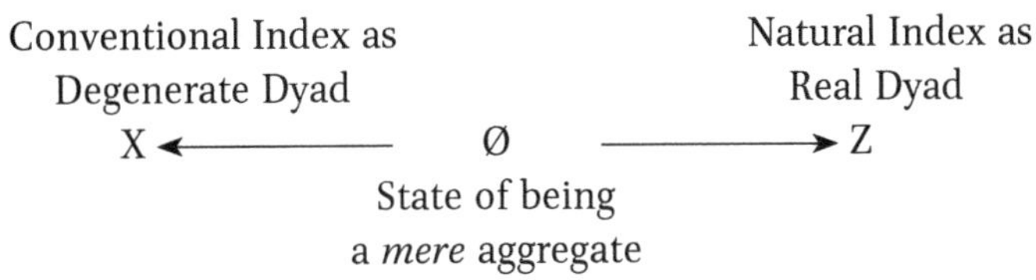

where both Z and X represent the result of negation applied to the primitive, basic category of being a *mere* aggregate. In this way, points Z and X are equally construed as refutations of the state of being a mere aggregate; hence, as designations of a state of *not* being or *non*-being a mere aggregate. Having in mind, now, that point Z expresses nothing but the essence of the genuine character of the natural sign (Index), which is causality, how do we then arrive at the explanation of what makes up the essence of the genuine character of the conventional index?

Starting now from point Z and moving back to Ø and further to X we may then see that the state of a genuine sign in X can only be arrived at by negating the causality of Z and thus moving to the causally unordered initial state Ø and, then, via Ø further to X by applying another negation.

Representing the causal character of the referential link *a* in Z as

(5.51) $CAUSALITY\,(a)$,

we then obtain the characterization of state Ø by the straightforward negation of (5.51), as it appears in (5.52):

(5.52) $NEG.CAUSALITY\,(a)$.

Strictly speaking, however, (5.52) will only be the designation of an operation, the latter implying that by the application of this operation state Ø is also obtained:

(5.53) $NEG.CAUSALITY\,(a) \rightarrow \overline{CAUSALITY}\,(a)$, the description of which in (5.53) appears as the consequent of the conditional. The opposite direction of negation applied to the state of being a mere aggregate as its starting point in Fig. III will then be represented as:

(5.54) $NEG.\overline{CAUSALITY}\ (a)$

As opposed to 'genuine indices' or 'reagents', when causality comes as the result of negating the negation itself in *NEG.CAUSALITY* (*a*), as is shown by means of the square brackets in (5.55):

(5.55) $NEG.NEG[CAUSALITY\ (a)] \rightarrow CAUSALITY(a),$

in the case of 'degenerate indices' (2.283), which Peirce also referred to as 'designations' (8.368n23), the negation of the random character of pairing in Ø does not appear as the result, but merely as the process of its attainment—in the same way as this distinction is presented by the opposition between the grammatical meaning of the Perfect and the Continuous aspects of the verb in English. Symbolically, this difference can be represented in terms of the opposition between (5.55) and (5.56):

(5.55) $NEG.NEG[CAUSALITY\ (a)] \rightarrow CAUSALITY(a),$
(5.56) $NEG.[NEG.CAUSALITY\ (a)] \rightarrow NEG.\overline{CAUSALITY}\ (a)$

In (5.56), it shows itself as the difference in the scope of the leftmost negation, which goes on a par with different ways of construing states X ($NEG.\overline{CAUSALITY}\ (a)$) and Z (*CAUSALITY* (*a*)), respectively.

3.6. *Negation, Mind Independence, and the Tone/Token/Type Distinction*

Preempting negation as a necessary element of procuring mind independence, which is obvious in the case of natural indices or 'reagents' (state Z in Fig. IV),[56] it will be seen that in the case of conventional indices (or else 'designations' and thus *non*-reagents) the respective 'derivative/secondary' causality required for the suppression of the unduly degenerateness of their primary sign-object pairing (that warrants the case of mind inde-

[56] ... due to the fact that in Peirce's genuine dyad the sign-object pairing is formed by natural causation alone, which is exactly what renders the denial of the chaotic character of the sign-object link in Z as mind independent.

pendence in state X) appears to be an intrinsic property of what would be defined (to borrow the term of chaos theory) as a state of *deterministic chaos* (see Fig. IV). Notably, this will be equivalent to defining state X as a dual of state Z, as the latter will then present itself—due to its straightforwardly deterministic character of causation—as a manifestly *deterministic* state:

Fig. IV.

Conventional Indices (*Non*-Reagents) X ⟵ Ø ⟶ Z Natural Indices (Reagents)

State of being a *mere* aggregate

State of deterministic chaos – State of (absolute) chaos – **Deterministic state**

Before pursuing the matter directly, some general remarks on the use of the term 'chaos' are in order. Thus, as Steven Jackett points out:

> The study of Chaos Theory is an important step in the gradual realization that things are considerably more complicated than a classical Newtonian deterministic interpretation, but contrary to what the word "chaos" might suggest, they are not completely random without any kind of underlying pattern or order for us to study. The word "chaos" was originally a misnomer because there really is underlying pattern and order for us to study, but it is complicated; we are changing the meaning of the word "chaos".[57]

Therefore, what the use of the term "chaos" in collocation with "deterministic" here directly implies is that the 'designative', "arbitrary", or 'non-reagentive' causality of state X, as a property of what we describe by the term *deterministic chaos*, differs from the classical Newtonian deterministic interpretation of causation in state Z exactly in the degree of the *internal complexity* of its deterministic interpretation. The underlying pattern, in this case, is merely much more vexed than that of the classical Newtonian deterministic interpretation of causation—and may thus look outwardly as being *non*-reagentive. However, in spite of what the misleading term "'non-reagentive' causality" might erroneously suggest, both

---

[57] Steven Jackett, *Philosophical Musings on The Implications of Chaos Theory on the Human Understanding of the Natural World.*

kinds of causality constitute basically one causal frame only differing in its ramifications, of which the case of designations turns up to be exceptionally hard to be discerned all too readily.

The latter fact can however be laid bare by taking into account yet further patterns of duality, which bring into play the respective Interpretants in the mind involved in sign interpretation. The latter duality renders itself as bearing on the known duality of ∀ and ∃, whereas the duality pattern in which Interpretants get involved appears in an intimate and purely technical correlation with the duality of '$p \wedge q$' and '$p \vee q$'. In particular this latter circumstance gives grounds to regard the respective Quasi-mind as having two kinds of mind *qua* transcendental *duals* as necessary parts of its transcendental constitution.[58] Speaking in the non-technical sense in which duality is understood as a systematic *reversal of* all *interpretations* (of which the reversal of *truth-functional values* happens to be just one of many instances),[59] the two kinds of mind in question, which we would dub, tentatively, as '*private*' mind, on the one hand, and '*public*' mind, on the other, will then be defined as duals of one another in the very same manner as ∃ is defined as a dual of ∀ and, conversely, ∀, as a dual of ∃. Accordingly, exactly the same kind of duality can also be noted between the pertinent Interpretants of the designation, the latter depending on whether the designation in question renders itself as a sort of an 'upgraded' Tone (i.e. as Second *qua* the result of its being sort of 'upgraded' from First) or else as something that might be described as a 'downgraded' Type (i.e. as Second *qua* Peirce's *Replica* of the corresponding Third).

In order to see, in further detail, how all this stands to the underlying nature of '*non*-reagentive' causality, notice that the *negation* of state ∅ (i.e., the *denial* of being a mere aggregate) may take two opposite directions on plane XZ (see Fig. IV). Thus, the link of state Z (which we choose as the starting point since state Z appears as basic and, thus, most primitive to the general notion of mind independence) to state X and the (reflexive) link of state Z to itself (similarly *via* state ∅, that of the initial chaos of being a mere aggregate) differ in respect of the direction of the negation

[58] Mind is, of course, different from brain, but this does not exclude that, between the two, there might be certain similarities of some abstract kind. Thus, human brain consists of two dual parts which are brain hemispheres. The same dual division pertains also to man's senses such as sight and hearing (and, in a rudimentary form, smell, too), also to paired organs such as kidneys etc. Even human population as such is constituted by physiological duals (male vs. female). There would be nothing unnatural in suggesting, then, that some or another sort of duality might turn out to be a basic characteristic of mind, too.

[59] Cf.: W. V. Quine, *Methods of Logic*, p. 69.

(or denial) of the state of absolute chaos (state ∅). The difference in question is visualized, in Fig. V, by the two-directional arrow just above the point of bifurcation in ∅, showing the part of the reflexive link of state Z to itself presented as line ∅ - Z/Z' as the mirror reflection of line ∅ - X:

Fig. V.

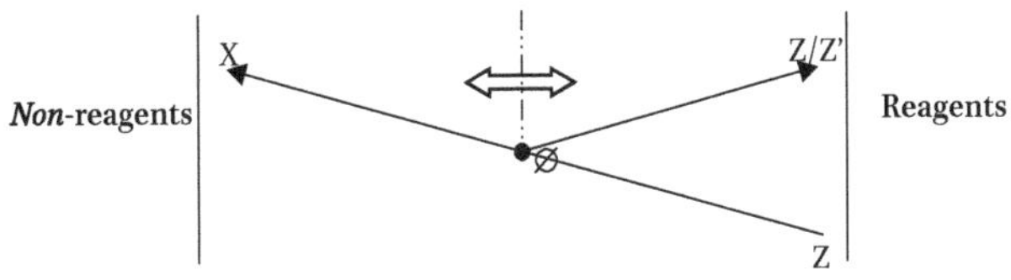

State of deterministic chaos – State of (absolute) chaos – Deterministic state

As a corollary, both of state Z/Z' and state X present themselves, on equal grounds, as duals not only in respect to one another, but also as subspecific duals of the deterministic interpretation of causation (Z) as such. And it is fairly evident that, being such subspecific duals, either of these states presents itself as a different kind of dual. Thus, as in the case of $\bar{p}$ which is a simple negation in propositional calculus that takes the same $\bar{p}$ as its dual,[60] state Z in a quite analogous way, as the negation of state Ø, turns out to be a *self*-dual in:

(6.61) $\mathit{CAUSALITY}\,(a) \rightarrow \mathit{NEG.NEG.}[\mathit{CAUSALITY}\,(a)] \rightarrow \mathit{CAUSALITY}\,(a)$,

where *CAUSALITY* (a), conceived in terms of being a plain negation of chaos, happens to be the same kind of self-dual as $\bar{p}$. On the other hand, since state X, as opposed to state Z', implies a non-rollback negation, it renders as a dual of an entirely *different* kind:

(6.62) $\mathit{CAUSALITY}\,(a) \rightarrow \mathit{NEG.}[\mathit{NEG.CAUSALITY}\,(a)] \rightarrow \mathit{NEG.}\overline{\mathit{CAUSALITY}}\,(a)$.

This time *CAUSALITY* (a) and $\mathit{NEG.}\overline{\mathit{CAUSALITY}}$ (a) appear to be dual to one another in much the same way as the logical content of, say, '$p \vee q$' happens to be dual to the logical content of '$p \wedge q$', but not to that of '$p \vee q$' itself.

Notice, however, that in the case of the duals under consideration, i.e. *CAUSALITY* (a) as a self-dual in (6.61) as well as *CAUSALITY* (a) and $\mathit{NEG.}\overline{\mathit{CAUSALITY}}$ (a) as duals of one another in (6.62), we are dealing with

[60] Cf.: W. V. Quine, *Methods of Logic*, p. 68.

something other than propositions like p or q as in our examples of '$p \vee q$' and '$p \wedge q$' as duals, and thus with such expressions that are not fully discernible in terms of the calculus of propositions. What we really need, then, is to look for more helpful analogies from elsewhere. The required analogy would not be hard to find, though, since the solution suggests itself immediately upon a careful look at the outward form of the expressions like (6.61)–(6.62). For what we have in *CAUSALITY* (a) as the consequent of the implication in (6.61) is but a functional expression, namely, an instance of the first-order function and, thus, an instance of the function from *objects to truth-values*. In contradistinction to *CAUSALITY* (a) as the consequent of the implication in (6.61), however, *NEG.*$\overline{CAUSALITY}$ (a) as the respective consequent of the parallel implication in (6.62) is also a functional expression but already of an entirely different kind. Namely, as a truth-function, *NEG.*$\overline{CAUSALITY}$ (a) is evidently an instance of the *second-order* function in which $\overline{CAUSALITY}$ (a) as a whole stands for its (first-order functional) argument. Thus, what is here illustrated in terms of first-order functional calculus by means of *NEG.*$\overline{CAUSALITY}$ (a) is in fact none other than an instance of Frege's early expression of generality, with *CAUSALITY* (x) as the sign of the first-order functional expression in question:

(6.63) $\vdash\!\!\top\!\!\smile^{\mathfrak{x}}\!\!\top\!\!-$ *CAUSALITY* (x)

Translation of (6.63) into modern notational convention will yield, likewise, the quantifier expression in:

(6.64) $\sim \forall x \sim$ *CAUSALITY* (x)

Since, for Frege, the expression '$F(x)$' of which *CAUSALITY* (a) is just an instance is none other than simply a kind of abbreviation for

(6.65) $\vdash\!\!-\!\!\smile^{\mathfrak{a}}\!\!-$ $F(a)$,[61]

which allows rendering, in modern notation, as:

(6.66) $\forall x F(x)$,

[61] According to Frege's rule of returning from an italic letter to a gothic letter with a quantifier, G. Frege, *Begriffsschrift*, §11; cf. also: W. Kneale and M. Kneale, *The development of logic*, p. 489.

the nature of the duality of $\mathit{NEG.}\overline{\mathit{CAUSALITY}}\ (a)$ to *CAUSALITY* (a) and, thus, of the duality of state X to state Z, becomes evident. For the simple reason that the duality in question turns out to be no more than a particular case of the logical duality of $\sim \forall x \sim F(x)$ to $\forall x F(x)$, with the particular instance of the function in question being quintessentially *CAUSALITY.*

For our purposes the logical conception of duality would not suffice, though, if we only were to restrict ourselves to its analysis in terms of the modern quantifier notation alone. Since what is at issue here is not only the duality of state X to state Z, but also the dual nature and thus the inner duality of the state of deterministic chaos itself, it would be convenient to go on by skipping the received quantifier notation in (6.64) and focusing, instead, on what has been inadvertedly lost in the modern convention of quantifier notation, and can thus be only explained by means of the original Fregean notation of generality in (6.63):

(6.63) $\vdash\!\!\!\top\!\!\!\smile^{\!\mathfrak{x}}\!\!\!\top$ *CAUSALITY* (x)

In contradistinction to the modern quantifier symbolism, which is employed in (6.64), Frege's use of the two content-strokes to the left and to the right of the concavity in the kind of expressions like (6.63) above as well as his rigid discrimination between these two strokes will be highly expedient for the purposes of the analysis at hand. Since the notion of '*CAUSALITY*' which we restrict to the description of the truly Newtonian deterministic state means or implies, rather, the most general property of natural cause-effect relationships, that of mind-independence, then the negation stroke attached to the horizontal stroke appearing to the *right* of the concavity and thus expressing the *particular* logical content of *CAUSALITY* (a) would evidently mean but the statement to the effect that in the particular occurrence under consideration mind-independence is definitely not the case. In other words, what it plainly indicates is that, in this case, the referential link between any of the two objects or phenomena involved in the sign-object schemata is clearly *mind-dependent*; hence, no strictly Newtonian cause-effect relationship can be at issue here. Notice, however, that the concurrent or parallel negation of the stroke of the logical content to the left of the concavity—quite contrary to what might be expected on the analogy with (6.61)—cannot serve as a negation of the negation of this particular case to the effect of stating its *mind independence*. For this simply follows as a corollary from the statement, made with the use of this negation, that it is not the *general* case that *CAUSALITY* (a) is mind dependent (since what is negated by the second negation stroke is not the particular but the general logical content of *CAUSALITY* (a)). *Nor, however, will it as well follow that some other such particular case could also be found.* For the statement to the effect that the negation of *CAUSALITY* (a) cannot be generalized takes an entirely different and quite peculiar sense here, which rules out the mere possibility of finding any other such *particular* case whatsoever.

The peculiarity of the case under consideration lies, as a matter of fact, in that the brute physical sense of *CAUSALITY* (a) as our particular instance of the logical function in $F(x)$ (where x stands for any ordered pair x_1, x_2 involved in the Newtonian relation of cause and effect) by itself prevents the very possibility of the received interpretation of the left-hand negation in (6.63). The traditional sense of the statement that "it is not the general case that $\sim F(a)$ is true" does not apply here for the simple reason that the Newtonian deterministic interpretation of causation in question that makes the precise sense of *CAUSALITY* (a) turns out to be no more nor less than just an *essential* property and thus the property of absolutely all the *f*'s under consideration, which goes utterly contrary to whatever assumption that it might however be found out to be the property of merely *some* of the *f*'s only.

As a corollary, in order for (6.63) to be true (which as a matter of fact it is), the effect to which its left-hand negation can only be used is that there is, indeed, at least or, rather, exactly one case in which $\sim F(a)$ is false. But, quite contrary to what the familiar quantifier idiom may suggest, the case in question cannot, hence, *does not* belong to *any of the particular cases of* $\sim F(a)$ in the sense of asserting that there can be at least or else exactly one of the *f*'s such that it is mind-independent. Just for this reason, namely for the reason that this case *cannot amount to any one of the particular cases* of $\sim F(a)$, the single legitimate option that remains is that the left-hand negation stroke, when asserted, asserts only about none other than the *general case itself.* Therefore, the necessary as well as sufficient condition for Frege's assertion stroke in (6.63) to express a true judgment is—or, in other words, the logical content of (6.63) can only be true—if, and only if, the left-hand negation stroke (that of the negation of the general content of the functional expression in question) relates solely to the general logical content *and thus does not relate to* any of the particular logical content of the functional expression at hand. But then—what follows from this as a corollary—if there is such circumstance under which the case *could* be mind-independent then it can only be the one in which the mind in question is not identical to, and is thus totally different from any of the minds, the dependence/independence with regard to which is being considered in *any* of the *particular* cases of $\sim F(a)$.

Thus we come, by means of a transcendental deduction, to acknowledging a purely logical necessity of stipulating a very specific kind of mind, namely *general* or else *'public'* mind, which is other than any one of the kind that we can find to be directly related to the particular, individual cases of $\sim F(a)$ and thus the type of mind that stands in the relation of a dual to the particular, individual mind in question.[62] The stipulation of the notion of public mind along with the notion of individual and thus, in a sense, "private" mind,[63] besides providing the only exegetical alternative, has also the advantage of lending powerful specific tools for the investigation into the nature of the referential link and, thus, giving us the expedient theoretical apparatus for providing yet further exegesis (somewhat over and above such that can be found in the literature like the one in terms of Kripke's explanation of ordinary proper names as *rigid designators*) as to the real mechanism of the dual sort of causality pertaining to the constitutive nature of the referential link in the case of conventional signs. Just as in the case of Saussure's semiological explanation of the total meaning of a linguistic sign in terms of his purely relational notion of *valeur*, the notion of duality provides us this time with convenient tools for the same type of relational explanation of causality and mind independence alike. Now it appears that these two minds must in an important sense be independent of each other, since the referential links found to be mediated by the general type of mind (which, if considered in strictly pragmatic terms, turns out to be none other than 'public' mind) will be found to be independent of the mind pertaining to the interpretation of the particular cases of $\sim$ *CAUSALITY* (a) as represented by the right-hand content stroke in (6.63), whereas the latter, in their turn, will be found to be independent of the general or else 'public' mind as the one that is in charge of the interpretation of the left-hand stroke and thus of the general logical content of '$\sim \forall x \sim$ *CAUSALITY* (x)' (see Fig. VI below):

[62] Cf. also A. Mołczanow, Quantification and Inference, *The Monist* 85:4, pp. 546–47.

[63] Namely, in the sense in which it may be shown to be directly associated with what Wittgenstein assumes under his term of 'private language', see Garver (1959).

Fig. VI.

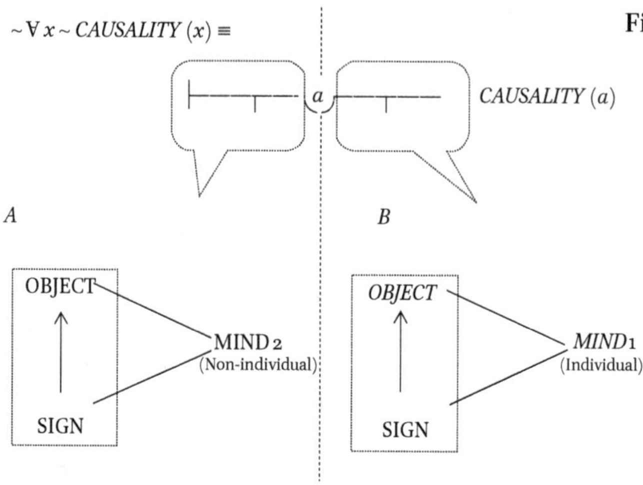

Under these circumstances scope relations will not seem to hold any more, since it will be quite evident that the negation of the particular logical content of *CAUSALITY* (a), i.e. the right-hand negation stroke in (6.63) will no longer be falling within the scope of the left-hand negation stroke:

(6.63) 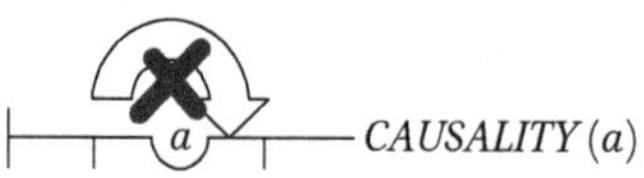

even though it might at first sight appear to follow from the linear order of negations in the Peano-Russellian convention of quantifier symbolism as represented in:

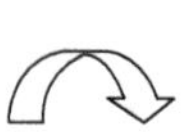

(6.64) $\sim\forall x \sim CAUSALITY(x)$

Notice that in the case under consideration the scope relations between the negations are identical with the respective scope relations that arise between the quantifiers themselves in cases like (6.67) or (6.68) below:

(6.67) $(\exists y)(\forall x)F(x, y)$
(6.68) $(\forall x)(\forall y)F(x, y)$[64]

---

[64] Cf. A. Mołczanow, *Logical Entailment on Multidimensional Branching Quantifier Representations*, pp. 133–34.

Due to the phenomenon of the so-called quantifier independence, such cases defy the linear ordering of quantifiers in conventional representations like (6.67):

(6.67) $(\exists y)(\forall x)F(x, y)$

The preferred symbolism used in the FPO (finite partially-ordered) quantification theory for the representation of such cases is widely known to be that of branching quantifiers[65] as shown in (6.67')–(6.68'):

(6.67') $(\forall x)$

$\quad\quad\quad\quad \searrow$

$\quad\quad\quad\quad\quad F(x, y)$

$\quad\quad\quad\quad \nearrow$

$\quad\quad (\exists y)$

(6.68') $(\forall x)$

$\quad\quad\quad\quad \searrow$

$\quad\quad\quad\quad\quad F(x, y)$

$\quad\quad\quad\quad \nearrow$

$\quad\quad (\forall y)$

In virtue of this fact, the scope relations between the negations in question would then be more adequately represented not by their "linear" Frege-style representation in (6.63) but, rather, by something like a respective "branching" version of this representation in (6.69), in which the concavity (i.e., Frege's original sign of generality) would be placed on a vertical line. Frege's original sign of generality would thus be clearly seen as none other than a mere divider between the (this time) vertical strokes of the general *resp.* particular logical content of the functional expression in question:

[65] Branching quantifiers were introduced by Leon Henkin, "Some Remarks on Infinitely Long Formulas", in *Infinitistic Methods,* Warsaw, 1959, pp. 167–83. For their theory, see e.g. W. Walkoe, "Finite Partially Ordered Quantification", in *Journal of Symbolic Logic*, vol. 35 (1970), pp. 535–50; Herbert Enderton, "Finite Partially-Ordered Quantifiers", in *Zeitschrift für Mathematische Logik und Grundlagen der Mathematik*, vol. 16 (1970), pp. 393–97; Jon Barwise, "Some Applications of Henkin Quantifiers", in *Israel Journal of Mathematics* 25, 1976, pp. 47–80; and Michał Krynicki and Marcin Mostowski, "Henkin Quantifiers", in *Quantifiers: Logics, Models and Computation*, Dordrecht, 1990, pp. 193–262.

(6.69)

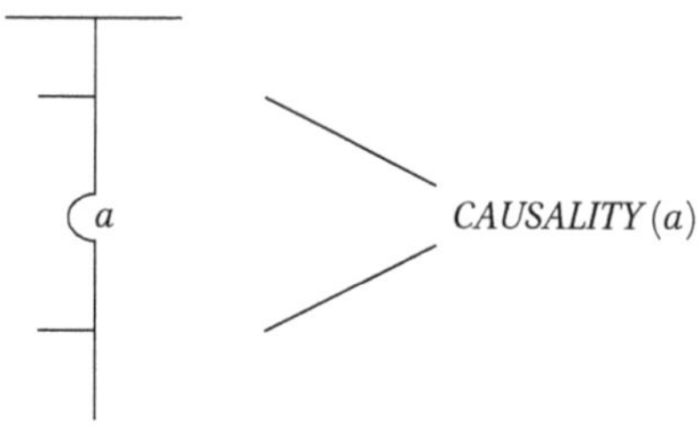

It is instructive to note that the use by Frege of his term "expression of generality" as the name of the concavity was evidently a misnomer even though, basically, the introduction by him of the concavity as such into his symbolism was what would now be seen, rather, as a step in the right direction. For, in strict terms, what really served the purpose of expressing generality was precisely the horizontal stroke of general content itself, but not the concavity as such. The factual importance of Frege's concavity thus boiled down to what it was doing purely graphically, *viz.* to providing the crucial division and thus making the distinction between the entirely different kinds of horizontal stroke. The stroke to the left of the concavity served as the symbol of the *general logical content* of the functional expression (or, for that matter, of whatever expression in general), whereas the stroke to the right, of the *particular logical content* of the very same expression. The real significance of Frege's innovation was thus not in the mere introduction by him of the concavity as the sign for expressing generality. Rather, it was only the consequence of his awareness of the necessity and of the logical importance of the radical discrimination between the general content and the non-general one on exactly the same grounds as it was carried out, independently of Frege, by Peirce through his distinction between Legisign and Sinsign, or else Type and Token. For the distinction between the strokes to the left of Frege's concavity and to the right of it was, basically, that between Legisign and Sinsign as such. Furthermore, Frege's left-hand and right-hand strokes were in fact signs that he devised in order to make explicit the logical relevance of what he saw as the deep contrast that arises between what he saw literally as the *two different sides* of the logical content of *the same expression* to which they both were to be prefixed. And this could only come out as the result of his being deeply cognizant of the fact that it was the very same expression that functioned as both a Sinsign and a Legisign in one. What could indeed be counted as being of most genuine importance in Frege's fundamental thinking underlying what seemed as a purely technical innovation appearing in the symbolic nomenclature of his concept-script was thus not only the discernment of the two kinds of logical object corresponding to Peirce's division into Sinsigns and Legisigns. For what was truly innovative in Frege's thinking could be summed up, using Peirce's terms, as this: the distinction between Sinsign and Legisign is not the distinction between the names of just two different *kinds of sign*; rather, it is the distinction between the names of the two different *kinds of content* of one and the same sign. Put another way, it is not only that the distinction between Sinsign and Legisign is logically important but, first and foremost, that *any sign is both* Sinsign *and* Legisign *in one.*

This latter point as well as the intricacies involved in the interplay of our dual negations can be better understood if we abandon the archaic representation in (6.69) and, instead, on the analogy with the Henkin-type branching quantifier configuration in (6.67')–(6.68'), represent the respective branching configuration of the negations in the way as it is done in (6.70) below:

(6.70)

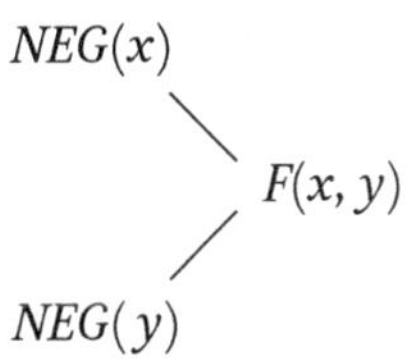

where $F(x, y) = \Phi(a)$,[66] $NEG(x)$ = the negation of the general logical content, and $NEG(y)$ = the respective negation of the particular logical content of $\Phi(a)$. In this case, contrary to the traditional interpretation under which $F(x, y)$ is normally understood as a function or relation with two arguments, it should nevertheless be regarded as no more than just an explicit representation of Frege's standard function $\Phi(a)$ with only a *single* argument-place. In this argument-place the letters 'x' and 'y' stand merely for a Fregean shift from a gothic letter to an italic letter depending on whether or not the functional expression with the letter in question appears with Frege's expression of generality as its prefix.[67] To put it a bit differently, the letter y in our representation of $F(x, y)$ in (6.70) stands for none other

[66] '$\Phi(a)$' appears here merely as an abbreviation for the representation of its particular case in '*CAUSALITY* (a)'

[67] ...and thus, more precisely, they stand for his shift from a gothic type to an italic type of *one and the same letter.*

than a certain specific value Δ of the argument of Frege's expression in Φ(Δ). The letter x is roughly equivalent to Frege's so-called gothic letter, which was Frege's own device (rather poorly understood by many of the numerous commentators of Frege's work, let alone completely abandoned or else totally ignored by any of his followers) merely to indicate that any such specific value for the argument just *cannot* be at issue here.

It would be illuminating to pursue the analogy between quantifiers and negations a bit further by regarding both the quantifier and the negation in the like manner, i.e. as a second-order function taking the value of the respective first-order function as the value for its argument. Since in this way the analogy turns out, as a matter of fact, to be none other than strict correspondence, one will hardly fail to see the point of branching configuration and, thus, of informational independence of the negations in (6.70). For what is at play, for example, in the case of informational independence of standard logical quantifiers ∃ and ∀ in (6.67'):

(6.67') $(\forall x)$

$\quad\searrow$ $F(x, y)$ $\nearrow$

$(\exists y)$

is the informational independence of the variable that is bound by ∃ from the choice of the variable bound by ∀ regardless of the linear order of these variables in their first-order function representation $F(x, y)$. More precisely, the dependence/independence relationships between the variables in cases like this are defined by FPO quantification theory in terms of the so-called *Henkin quantier*:

(6.71) $\forall x \exists x'$
$\forall y \exists y'$

This non-linear, or branching quantifier prefix represents dependencies between variables when quantifiers cannot be written down in a linear ordering. If we consider the ordinary universal and existential quantifiers appearing in (6.71), i.e. $\forall x$, $\forall y$, $\exists x'$, $\exists y'$, with the dependencies between the variables bound by these quantifiers defined as "for every x, y there are x' independent of y, and y' independent of x," then considering only the orderings in which $\exists x'$ appears after $\forall x$ and $\exists y'$ after $\forall y$, and ignoring inessential differences of ordering between the two universal quantifiers $\forall x$, $\forall y$ and of ordering between the two existential quantifiers $\exists x'$, $\exists y'$, we will only be left with the following three linear orderings:

(6.711) $\forall x \forall y \exists x' \exists y'$
(6.712) $\forall x \exists x' \forall y \exists y'$
(6.713) $\forall y \exists y' \forall x \exists x'$

In (6.711) and (6.712), however, the choice of y' turns out to depend on x, whereas in (6.713) it is the choice of x' that happens to depend on y.[68] So the Henkin quantifier (6.71) remains the only appropriate model to represent the dependencies in question.

The applicability of the FPO quantification theory for the representation of cases of informational independence like the one exemplified in (6.67') gives us the expedient means for the analysis of the equally branching relationships in (6.70):

(6.70)
$$\begin{array}{l} NEG(x) \\ \qquad\qquad \searrow \\ \qquad\qquad\qquad F(x, y) \\ \qquad\qquad \nearrow \\ NEG(y) \end{array}$$

Reducing (6.70) to a standard Henkin quantifier prefixed FPO quantification formula in (6.72):

(6.72)
$$\begin{array}{l} \forall x \exists x' \\ \qquad\qquad \searrow \\ \qquad\qquad\qquad \bar{F}(x, y, x', y') \\ \qquad\qquad \nearrow \\ \forall y \exists y' \end{array}$$

and further to:

[68] M. Krynicki, M. Mostowski, *Henkin Quantifiers*, p. 196.

(6.73)
$$\begin{array}{lll} \forall x \exists x' & & \\ & \diagdown & \\ & & S(x, y, x', y') \\ & \diagup & \\ \forall y \exists y' & & \end{array}$$

which we obtain as the result of substituting in (6.72) of S in place of $\bar{F}$ (as the negation of F), we see that in all the important respects (6.73) appears to be equivalent to our representation containing branching negations. For the semantic interpretation of (6.73) in terms of its skolemization in (6.74)–(6.75):

(6.74) $S(x, y, f(x), g(y))$
(6.75) $\exists f \exists g \forall x \forall y S(x, y, f(x), g(y))$

turns out to render the same sense as that of (6.70):

(6.70)
$$\begin{array}{lll} NEG(x) & & \\ & \diagdown & \\ & & F(x, y) \\ & \diagup & \\ NEG(y) & & \end{array}$$

and thus (6.70) happens to be recursively isomorphic with (6.73):

(6.73)
$$\begin{array}{lll} \forall x \exists x' & & \\ & \diagdown & \\ & & S(x, y, x', y') \\ & \diagup & \\ \forall y \exists y' & & \end{array}$$

due to the equivalence of the respective second-order interpretations. This follows naturally as a corollary of the duality of state X to state Z. Since in our case *NEG*(x) and *NEG*(y) are not simple but just dual negations, what is hidden in disguise of these dual negations are none other than the respective choice functions on the domains of 'x' and 'y'. Just owing to the dual character of the negations involved, these domains form strictly disjoint subsets of set A, that of *Non-F*'s:

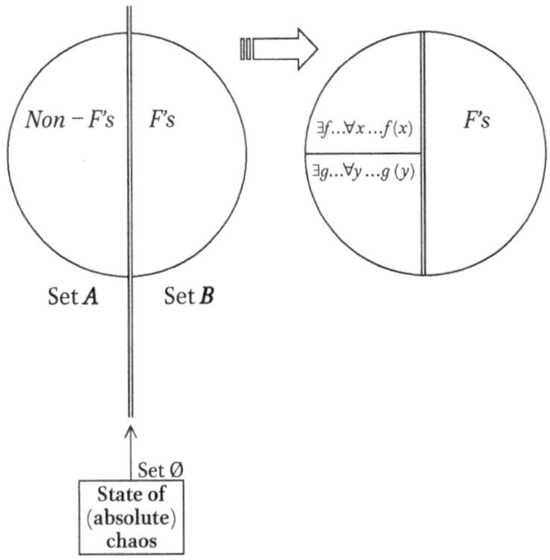

Fig. VII.

The choice functions standing behind the branched negations and shown to the right of the arrow in Fig. VII as $f(x)$ and $g(y)$ in $\exists f \ldots \forall x \ldots f(x)$ and $\exists g \ldots \forall y \ldots g(y)$ operating on these disjoint subsets as their domains are as a matter of fact the same Skolem functions introduced in (6.74)–(6.75) by skolemization of (6.73) relative to the Henkin quantifier in (6.71). The function symbols $f(x)$ and $g(y)$ that we get as the result of skolemization by substituting either of the occurrences of x' and y' in $S(x, y, x', y')$ are peculiar in the sense that, in our case, they indicate functions of a very specific sort. To see this, consider one of the simplest forms of the Axiom of Choice:

Axiom of Choice. Let C be a collection of nonempty sets. Then we can *choose* a member from each set in that collection. In other words, there exists a function f defined on C with the property that, for each set S in the collection, $f(S)$ is a member of S.

Taking the disjoint subsets of *Non-F*'s as such collections, we will see that in our specific case the choice function takes on a very special role. That is, the instances of x' that $f(x)$ will be seen as picking up turn out to be none other than what in Peirce's terms could be described as *Replicas* of a given *Type*. The relation between *Type* and its *Replica* would thus be seen as the substance of the $\forall x \exists x'$ prefix, i.e. of the upper part of the Henkin quantifier in (6.71) above, which we reproduce here as (6.76):

(6.76) $\forall x \exists x'$
$\forall y \exists y'$

where the universal quantifier $\forall x$ and the existential quantifier $\exists x'$ would stand for *Type* and its *Replica*, respectively. On the other hand, the respective instances of y' picked up by the second choice function, $g(y)$, and the functional relation itself would then make up the substance of the lower hand relation in (6.76), which is the relation between the universal quantifier $\forall y$ and the existential quantifier $\exists y'$. In the latter case, the relation between the universal quantifier $\forall y$ and the existential quantifier $\exists y'$ will be strictly equivalent to that between Peirce's *Tone* and *Token*. Notice now, that what (6.76) shows is that the choice of *Replica* will be dependent on the choice of *Type*, whereas the choice of *Token*, on that of *Tone*. This is fully exhibited in Fig. VIII:

Fig. VIII.

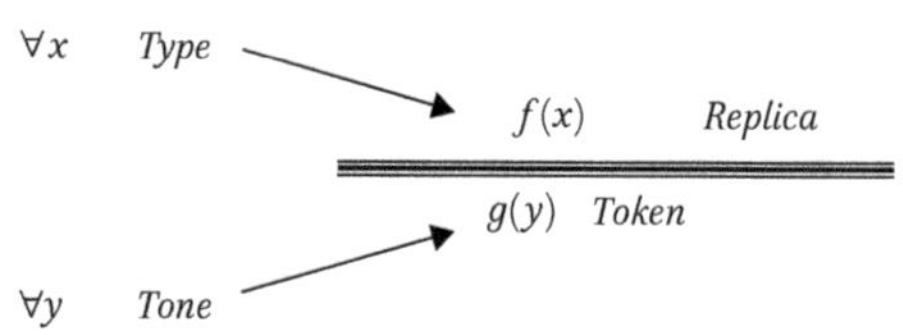

where *Type* is shown as the domain element of $f(x)$, and *Replica* as its range element, whereas *Tone*—as the domain element of $g(y)$, and *Token*, as the range element of $g(y)$.

Rewriting now '$F(x)$' (which we used as just another way of symbolizing '*CAUSALITY*(a)' that serves as the representation of natural cause-effect relationships of deterministic state Z) as '$F(z)$', and comparing it with the skolemization in $S(x, y, f(x), g(y))$:

(6.77) $S(x, y, f(x), g(y)) \backslash F(z)$,[69]

we will see that the obtained formula (where "\" is read as the sign of exclusion) gives us the full-fledged formal representation of the nature of the secondary, or derivative causality of state X (state of deterministic chaos). Since the secondary, or derivative causality in question makes

[69] ...where S is an appropriate functional abbreviation for our symbolization of the causality of state X in our expression in '*NEG.*$\overline{CAUSALITY}$ (a)'.

up the material content of the referential link in the case of an arbitrary (non-natural) sign, the formula in (6.77) thus provides the appropriate exegesis of the nature of the conventional sign as such. The full structure of the conventional sign S is then represented as the quadruple $x, y, f(x), g(y)$, where:

1. *The first pair* x, y represents a relation between what we would term tentatively as a "conceptual" facet and a physical facet of sign S;[70]
2. *Relation* $x \ldots f(x)$ renders an instantiation of what would traditionally, though erroneously (see Section 4.1 below), be regarded as Peirce's Third (which is *Type* symbolised as 'x') in a corresponding Second, namely in the *Type*'s *Replica*, the latter having its counterpart in the value of the correlated Skolem function;
3. *Relation* $y \ldots g(y)$ represents an instantiation of Peirce's First (namely *tone* in y) in a corresponding Second, which is *token* represented as the value of the respective Skolem function;
4. *The two Skolem functions* the existence of which is laid down by the quantified skolemization of (6.74) in $\exists f \exists g \forall x \forall y S(x, y, f(x), g(y))$ epitomize the causality of conventional signs as the pertinent counterpart of the physical causality of natural indices.

What is now also easily explained away as a natural corollary of the former kind of causality is *mind independence*, the latter fully exhibited by

5. *The relation of mutual independence between* x *and* $g(y)$ along with *the relation of mutual independence between* y *and* $f(x)$.

The more complete in depth layout of the last and the most important of the relations in question—which is Peirce's Third as a relation between two Seconds and at the same time the one expressing the essence of quantification—namely the respective

6. *Relation between the values of the Skolem functions,* $f(x)$ *and* $g(y)$ will follow in the sequel.

---

[70] ... this relation happens to be on a par with F. de Saussure's somewhat similar relation between the 'signifié' and the 'signifiant', the most significant difference being that in Saussure's interpretation the 'signifiant' was however a psychic phenomenon, a *sound-image*, but not the physical sound *per se* (F. de Saussure, *Course in General Linguistics*, p. 67).

CHAPTER FOUR

PEIRCE'S CATEGORIES AND THE TRANSCENDENTAL LOGIC OF QUANTIFICATION

> These philosophical discussions speak to issues central to linguistic theory, they raise questions about the nature of meaning and sameness of meaning, the nature of reference and the nature of linguistic knowledge. Yet these issues are rarely expressed fully, nor made explicit. The type-token relation as it is commonly conceived rests on a naïve distinction between abstract objects and physical objects, and on the equation of meaning and significance with the abstract. The linguistic unit is divided between two modes of being: in as much as it is meaningful, it must be abstract. In as much as it is an object—ink marks, sound waves, etc.—it is a physical object. It emerges further that this naïve and unexamined distinction gives rise to one insurmountable problem, that of saying how these two sides of the ontological divide can be joined.
>
> *C. M. Hutton, Abstraction and Instance, p. 5*

> The lesson of the theory of understanding is principally that the real generative structures of understanding, conception, and thought belong to our existence, internal existence, but are largely hidden from consciousness. Within consciousness the objects of these things appear, informed with recognition and conceptual Gestalt; and while we have no trouble speaking of those objects, difficulties arise when we attempt to turn thought on its own internal operations.
>
> *K. L. Ross, The Origin of Value in a Transcendent Function*

> Human beings can cognize and know only either sensory appearances or the forms or structures of those appearances—such that sensory appearances are *token-identical* with the contents of our objective sensory cognitions, and such that the essential forms and structures of the appearances are *type-identical* with the representational forms or structures generated by our own cognitive faculties, especially the intuitional

> representations of space and time—and therefore we can neither cognize, nor scientifically know, nor ever empirically meaningfully assert or deny, anything about things-in-themselves.
>
> *Immanuel Kant, Critique of Pure Reason A369/B310–11, emphasis added*

4.1. *Degenerate Thirdness vs. Thirdness as Relationship*

As Jaime Nubiola mentions in his lecture on "Peirce on Complexity":

> The easiest access to the categories is by way of experience. All we need is to simply *look* at how *phenomena* appear. This is exactly what Peirce suggested by his choice of the word 'Phaneroscopy' (from the Greek words tó fanerón, which is synonymous to fainomenon, and skopein, which means 'to look at'). Such phaneroscopy "shows that the formal relations studied in mathematical logic have material correlates in experience" (Parker 1998: 105). Let us take an example of my feeling the solid surface of this desk: as feeling, it involves reactivity, opposition, and thus secondness. But how the two elements are related to each other so that there is the object which I call this desk, is a matter which Peirce calls Thirdness. On the other hand, since secondness somehow presupposes that there are two elements involved, each of which is distinct from the other, the phaneroscopy must admit of Firstness which, in virtue of its sheer singleness is the most difficult aspect to describe:
>
> Indeed, strictly speaking it cannot be described without contaminating it with an element of thirdness. Firstness is that element of an appearance which does not refer to anything other than itself. The closest we may come to describe firstness is by attempting to think a sensation before we sense it (Debrock 1996: 1339).[1]

No wonder that the contamination of Firstness with an element of Thirdness can hardly be avoided in the depiction of Peirce's *Type* as well. What can be highly instructive here is the passage from Peirce, where he attempts to distinguish the generality of Thirdness from that of Firstness through the example of a cook following a recipe:

> She [the cook] is directed to take apples. Many times she has seen things which were called apples, and has noticed their common quality. [...] What she desires is something of a given quality; what she has to take is this or that particular apple. From the nature of things, she cannot take the quality but must take the particular thing [...] She has only seen particular apples,

[1] Jaime Nubiola, "Peirce on Complexity". In W. Schmitz, ed., *Proceedings of the 7th International Congress of the IASS-AIS*, Thelem, Dresden, 2001, pp. 11–23.

> and can take only particular apples [...]. Throughout her whole proceedings she pursues an idea or dream without any particular thisness or thatness—or as we say, *hecceity* [= Secondness]—to it, but this dream she wishes to realize in connection with an object of experience [...]. The dream itself has no prominent thirdness; it is on the contrary utterly irresponsible [...]. The object of experience as a reality is a second. But the desire in seeking to attach the one to the other is a third, or medium. So it was with any law of nature. Were it but a mere idea unrealised—and it is of the nature of the idea—it would be a pure first. The cases to which it applies are seconds.[2]

Mutatis mutandis, the same can be said of Peirce's *Type*. Just as a mere idea unrealised would be a pure first, so would be a mere *Type*, too. Only the cases of its instantiation given in sensuous perception, i.e. *Replicas*, would clearly be Seconds. Likewise, what Peirce calls 'the desire in seeking to attach the one to the other' that he describes as 'a third, or medium', clearly translates as a synthesis which is not itself contained in perception, but which contains the synthetical unity of the manifold of perception in a consciousness, is presented by our Skolem function, $f(x)$. A similar medium, or third we can also find in the other of the Skolem functions in question, that of $g(y)$. But these, as we know, are *degenerate* thirds. Their degenerateness is quite apparent since they are both media appearing within *degenerate* dyads and, thus, what they mediate are matters of random choice, not matters of fact. The latter happens only in the case of the *real* dyad, constituted of two indisputable Seconds. Of the Skolem functions in question, however, either mediates between a sheer *First*, on the one hand, and only on the other, a true *Second*. Therefore, the *real*, full-fledged, untruncated *Thirdness* we can only find in our case in a relation between the Seconds themselves. This goes on a par with the relation between the respective two Seconds in a *real* Dyad, as in the case of 'natural' signs. And, most importantly, as also in the case that Pharies refers to as one of "the most important manifestations of Thirdness," namely reasoning, which "... is a triadic process of a syllogistic nature, in which an inference is drawn from the comparison of two facts, the inference acting *the two facts into a related whole*".[3] These two Seconds, or facts in question appearing in a related whole in $S(x, y, f(x), g(y)) \setminus F(z)$ as the *values* of our Skolem functions $f(x)$ and $g(y)$, are none other than the *Type*'s *Replica* and the *Token*, by the use of which either is regarded (in a relation to its referent) as a special case of *Index*.

[2] *CP* 1.341–2.

[3] David Pharies, *Charles S. Peirce and the Linguistic Sign*, p. 12.

Notice, however, that—as has already been pointed out in our discussion of the genuine role of Frege's concavity serving as a divider between the general and the particular content of the same expression—the *Replica* and the *Token* turn out to be the same sign considered *qua* Object. Or, to put it the other way round, any expression, or sign turns out to be both Sinsign and Legisign at the same time. What makes real difference in the relation between the values of $f(x)$ and $g(y)$, then, is not the relation between the Type's Replica and the Token (as they are the same sign), but the relation between the general and the particular content of this very sign. Taking the latter into account, it is easy to observe strict equivalence that holds between the material identity of the same sign both as a Type's Replica and as a Token on the one hand, and the formal identity of the same middle term acting as part of both premises of the syllogism, on the other (see Fig. IX):

Fig. IX.

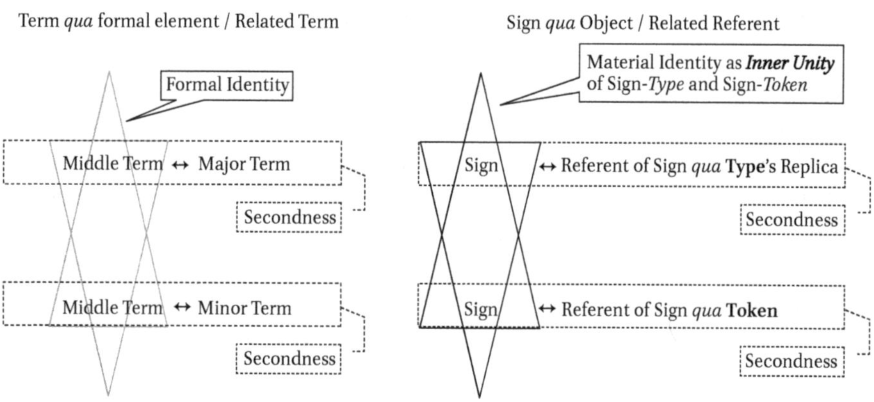

Fig. X.

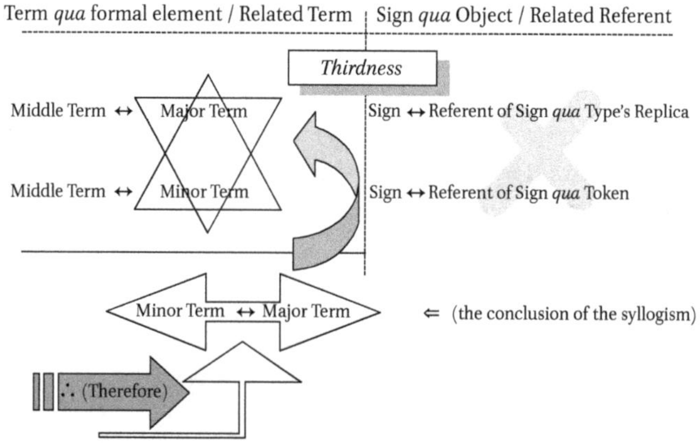

Therefore, in much the same way as it happens in syllogistic reasoning, in which the triadic character of Thirdness reveals itself as inference through the relation between two non-middle terms that gets explicitly exposed by the conclusion of the syllogism (Fig. X), Thirdness also manifests itself in a similar way through a triadic relation between the two referents of the sign making the essence of quantity that gets exposed in ordinary language.[4] Namely, it discloses itself through the respective relation between the referent of the sign *qua* Type's Replica and the referent of the same sign *qua* Token (Fig. XI):

Fig. XI.

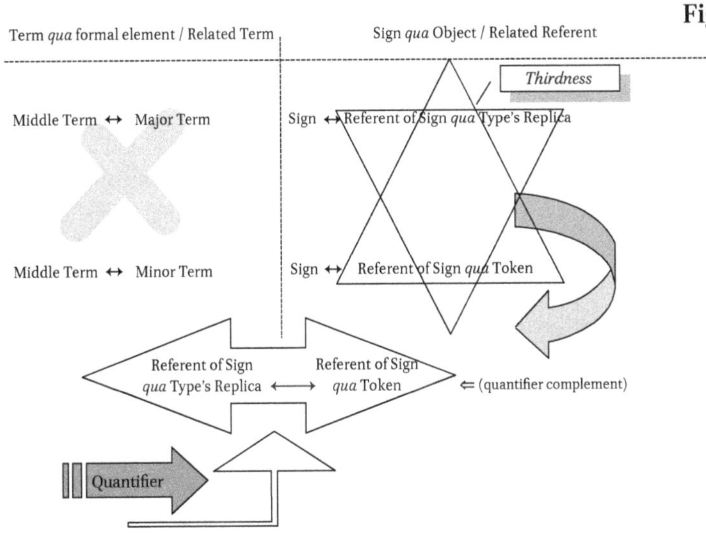

Accordingly, the question arises if there exists some explicit way of representing Thirdness that would be analogous to its formal representation that we find in syllogistic reasoning, where that role is taken over by the conclusion of the syllogism in which the two limiting terms are explicitly bound together by means of a copula. For this may constitute a "clue" to the proper discovery of the categories of understanding.

[4] Since it no longer makes any sense to speak of whatever exegetic value of the logical concept of the quantifier, we focus now solely on quantification in natural language as the only worthwhile object of a transcendental consideration, taking into due account that "language is fully included within the explanatory scope of Kant's general theory of objective mental representation... [which] comprehends non-linguistic and linguistic meaning alike" (Robert Hanna, *Kant and the Foundations of Analytic Philosophy*, p. 19).

The schemata shown in Fig. IX–XI happen to be of crucial relevance here, as they come to show explicitly the basic premise of Kant's Transcendental Deduction that resides in the assumption of an intrinsic and intimate connection between apperception and judgment. As Paul Guyer has most recently noted:

> The obscurity of both of Kant's versions of the Deduction . . . arises from the fact that in neither version does Kant . . . successfully exploit the "clue" to the discovery of the categories by clearly expounding the connection between apperception and *judgment*. More precisely, in the first-edition version of the Deduction, Kant omits any explicit account of the connection between apperception and judgment, while in the second edition, clearly having become aware of this problem in the interval, Kant burdens his argument with a problematic conception of judgment itself and an account of the connection between apperception and judgment that undercuts the original premise of the ubiquity of apperception itself.[5]

It is pretty evident that the striking parallelism that we observe in Fig. IX–XI may contain a suitable and fitting key to the discovery of the categories as well as show that the conditions of their validity are based on the very same principles that also warrant the validity of the conclusion.

In the expressions that can be found in natural language (as opposed to Frege's *Begriffsschrift*), any kind of analogous explicit syntactic juxtaposition of the Type's Replica and the respective Token seems, however, to be highly problematic, due to their purely material identity.[6] Yet there is another kind of juxtaposition that turns out to be highly relevant to the subject matter under discussion. In just the very same way as it is in a syllogistic structure where we are bound to find an indication of inference in the word 'therefore',[7] the respective indication of the Thirdness in question will be quite naturally found in a formal element that similarly occurs in juxtaposition to the signifying expression. In the medieval tradition, such formal elements were referred to as *syncategoremata*, and in the linguistic tradition after Fries, as *noun determiners*.

---

[5] P. Guyer, *The Deduction of the Categories: The Metaphysical and Transcendental Deductions*, p. 123.

[6] It will be shown in the sequel, though, that a special kind of explicit juxtaposition is nonetheless conceivable in the bordering case of the so-called *partitive* use of quantifiers, as opposed to the *non-partitive* use of quantifiers considered here (Jespersen 1969:118; Carden 1976:1), appearing in the wider context of the distinction between delimitative and juxtapositive serializations of English noun phrases (Thrane 1980: 126), where the role of the preposition 'of' with regard to Thirdness will be readily seen to be strictly akin to that of the copula.

[7] . . . or in its equivalents like English *hence*, Latin *ergo*, etc.

In the position of structural linguistics after Fries these formal elements such as articles and "quantifiers" (Jespersen's term for "words denoting quantity") were regarded as so-called structural signals, i.e. as words of formal classes that serve to signal structural (grammatical) meanings, as opposed to the notional, or lexical meanings expressed by the notional words with which they occur. But, amounting as their name suggests to just signalling that the element to which they are adjoined is a word that belongs to the notional class of nouns, their function happens to be the function of the plural ending of the noun as well. In other words, the same role of structural signal is also played by an inflectional morpheme that occurs (on a morphemic level) in juxtaposition with a notional, or root morpheme. Moreover, the latter when devoid of this structural signal becomes ambiguous in much the same manner as the conclusion of the syllogism itself acting *the two facts* (i.e. the two syllogistic premises) *into a related whole* when this conclusion contains no indication as to whether the conclusion is an *A*-statement or an *I*-statement. Exactly this happens in the case of the so-called *indefinite* or *indesignate* propositions like

(1.1) a. *Metals are useful*
b. *Comets are subject to the law of gravitation*

which are ambiguous between (1.2) and (1.3):

(1.2) All { Metals are useful / Comets are subject to the law of gravitation }

(1.3) Some { Metals are useful / Comets are subject to the law of gravitation }

Regarding such propositions, Jevons in his *Elementary Lessons in Logic* notes that

> We may safely take the preceding examples to mean "*some metals* are useful" and "*all comets* are subject to the law of gravitation", but not on logical grounds. Hence we may strike out of logic altogether the class of indefinite propositions, on the understanding that they must be rendered definite before we treat them.[8]

---

[8] W. Jevons, *Elementary Lessons in Logic*, p. 65.

It will be readily seen, in the light of Jevons' remark, that besides the A- or I-Quantity of propositions that require words like *all* or *some* to render them definite, what evidently needs to be rendered definite in the like manner is, quite apparently, also the respective Quantity of the notional, or root morpheme, and thus of the morphosyntactic *Thirdness* construed (see also Fig. XI above) as follows:

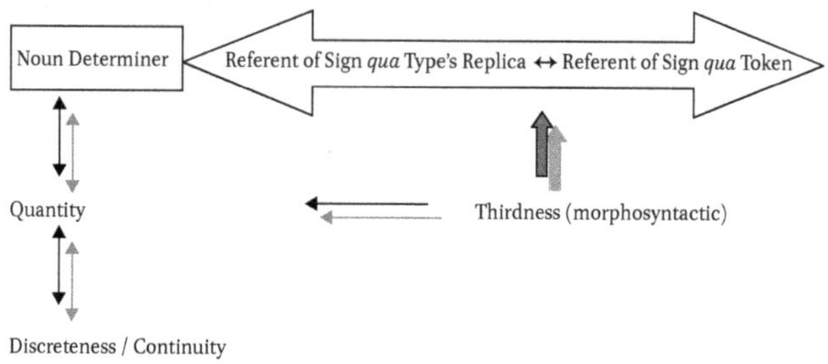

In the latter case, the most noticeable designations of the respective Quantity of Thirdness are clearly the various elements labelled collectively as noun determiners that appear in juxtaposition to the notional element representing Thirdness. Evidently, their function as noun determiners (i.e. their function as words that determine that the word juxtaposed is a noun) turns out to be a mere sequella to, or, so to speak, a collateral effect of, their primary function of 'signalling' Thirdness and, thus, only derivative to their function of designating Quantity as the basic characteristic, or property, of the respective morphosyntactic Thirdness. On the more basic morphemic level, their latter function comes, in its turn, to be crucially dependent on a purely morphemic function of designating the most basic characteristic of Quantity, namely *Discreteness* vs. *Continuity*.[9]

As a corollary, noun determiners (or 'word-morphemes') play the role of the designations of quantity as the basic characteristic of morphosyntactic Thirdness only in a supplementary way to that played by morphemes *per se, viz.* noun inflections that come out to designate discreteness or continuity. As morphemes, the inflections in question do belong to the internal morphological structure of the noun in much the same way as *some* and *all* fit in the internal syntactic structure of propositions in "[*some*] *metals*

[9] ... or, rather, *discontinuity* vs. *continuity*, which Fraenkel, Bar-Hillel and Levy refer to as "one of the oldest and most intricate notions of science in general": cf. Fraenkel A. A., Y. Bar-Hillel, A. Levy. *Foundations of Set Theory*, Ch. IV, § 1. Historical Introduction. The Abyss Between Discreteness and Continuity, p. 211.

are useful" or "[*all*] *comets* are subject to the law of gravitation." However, expressions like *some* and *all* acting as indicators of the respective quantity of propositions, appear to be in stark contrast to such expressions as *therefore, thus* and *hence* that do *not* properly belong to the structure of the proposition. Accordingly, noun inflections are to be found in the same type of contrast to word-morphemes (*viz.* noun determiners such as articles and quantifiers) that do not belong to the internal morphological structure of the noun. The role of the "internal" indicator of Thirdness at the level of sentence-syntax as well as word-morphology would then be seen as more basic and more fundamental than the role of the indicator that appears as "external" at either of these levels. The priority of the choice of words like *some* as "internal" adjuncts to *metals are useful* over the choice of *therefore* as an "external" adjunct to the whole of the proposition in *some metals are useful* is quite evident:

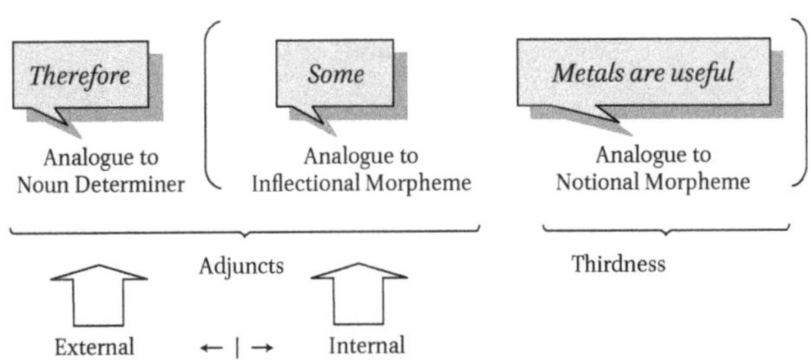

If we were to pursue the analogy between "syllogistic" and "morphosyntactic" Thirdness yet further, it would be worthwhile taking notice that the ambiguity in Jevon's examples (1.1a)–(1.1b) is being resolved, and thus the required 'definiteness' attained, simply by the indication of whether the major and the minor terms of the proposition are congruent or not. For this is exactly what *all* and *some* do by indicating, "whether the predicate is applicable to the whole, or only part of the subject."[10] Hence, what we may reasonably expect is that the function of the corresponding internal indicators of the "morphological" Thirdness will, too, amount to an indication of congruence, only this time in the basic properties, i.e. continuity or discontinuity, of the quantity of the referents in question. Thus, in Slavonic languages such as Polish or Russian the role of singular/plural noun endings '-a' and '-i' with regard to Thirdness will be seen as fully analogous to that of *some* and *all*:

[10] W. Jevons, *Elementary Lessons in Logic*, p. 65.

Fig. XII.

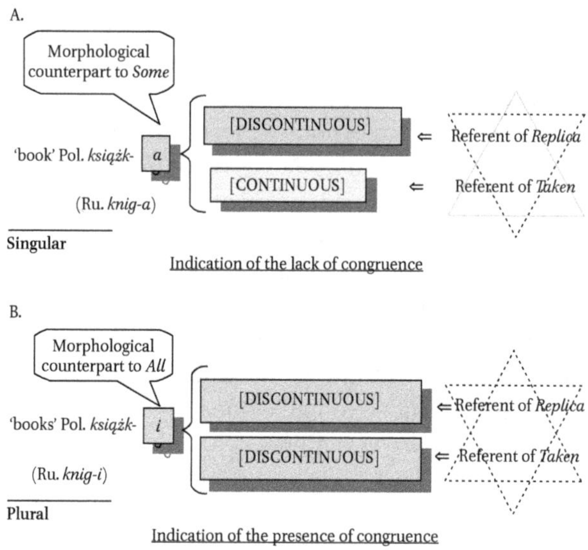

The parallel arises just in view of the fact that the plural ending '*-i*' in both Polish and Russian happens to indicate precisely the congruence in the discontinuity of both the referent of the notional morpheme *książk-/knig-* 'book' *qua* Replica and the referent of the same notional morpheme *qua* Token (Fig. XII B), whereas the singular ending '*-a*' shows none other than the obvious lack of such congruence (Fig. XII A). More specifically, in (XII A), similarly as in the syllogistic case of *some*, the quality of being 'continuous' of the referent of Token (the counterpart to the syllogistic predicate) is applicable only to the *part* of the 'discrete' referent of Replica (the counterpart to the subject), i.e., to the part that, as such, is 'non-discrete'. In (XII B), on the other hand, similarly as in the case of *all*, the quality of being 'discontinuous' of the referent of Token (≈ the predicate) is applicable to the *whole* of the similarly 'discontinuous' referent of Replica (≈ the subject).

Accordingly, both singular and plural noun endings may become redundant when such disambiguation is not required. This is exactly what happens in the case of the so-called uncountable nouns like Polish *śnieg* 'snow' or Russian *sneg* 'snow' (Fig. XII C):[11]

[11] These examples are intended for expository purposes only, and thus they need not fulfil the task of providing any exhaustive inventory of the category of number in Russian and Polish alike.

Fig. XII.

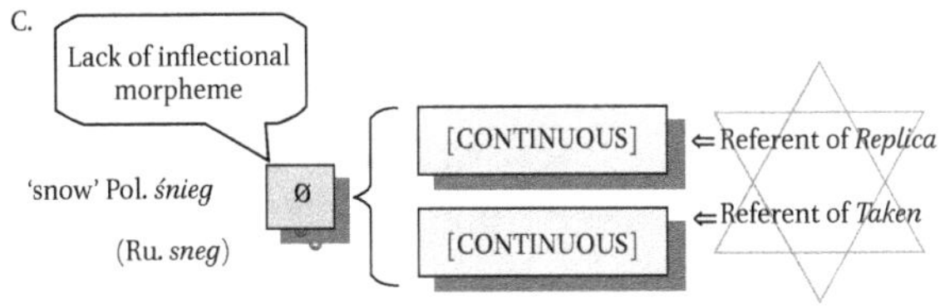

Note that these nouns owe their lack of an inflectional morpheme not to the obvious homogeneity, of the quantitative characteristics of Thirdness, as such (cf., e.g., the analogous case of *all* as well its morphological counterpart in Fig. XII B), but to the very fact that in this case the issue of the choice between homogeneity and heterogeneity simply does not arise.

The case with uncountable nouns as in Fig. XII (C) presents itself, however, a bit differently in English. The known fact that Modern English denies any specific non-zero ending for singular nouns makes—for general names like *book* and *snow* (in fact, for any general name)—the basic 'Referent of *Replica*/Referent of *Token*' ratio ambiguous between the readings of XII (A) and (C). Just to compensate for this lack of the non-zero ending, the required definiteness in this case needs to be attained by the additional use of an "external" indicator of the relational characteristics of Thirdness. In English this function of an "external" indicator is evidently performed by the indefinite article, which plays exactly the same role as that taken on by the plural ending '-*s*' of English nouns in resolving the parallel ambiguity between XII (A) and (B).

What is more, since the indefinite article in English shares its pre-nominal position with other noun determiners, it would be quite natural to expect that quantitative determiners in English, *viz.* English quantifiers, will take over the same function, that of disambiguating the respective relational characteristics of Thirdness. Notice, for instance, that due to the richness of the inflectional system in Russian, the need for such additional external disambiguation simply does not arise. Thus, the inventory of Russian quantifiers can do with just a single quantifier item occurring with both countable and uncountable nouns, e.g. ***malo*** *knig* ('few books'), ***malo*** *snega* ('little snow'); or else ***mnogo*** *knig* ('many books'), ***mnogo*** *snega* ('much snow'). In the latter case, the same quantifier item, *mnogo*, occurs regardless of whether the following noun is countable or uncountable:

Fig. XIII.

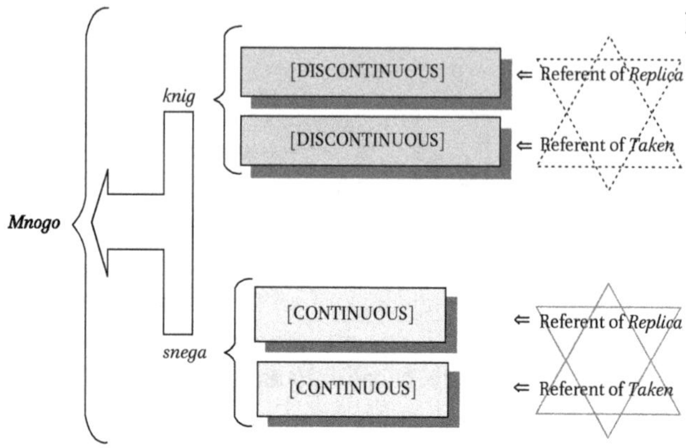

In English, however, the analogous use of the same undifferentiated quantifier item in the article position will not do for the simple reason that the relevant difference in the relational characteristics that in the absence of quantifiers is marked by the occurrence or non-occurrence of *'a/an'* before the noun will then be completely obliterated. This clearly indicates that the distinction between the two "synonymous" quantifier items (either of which is translated as the same quantifier in Russian) serves as a complementary means of disambiguation of the relational characteristics of Thirdness in the underlying semiotic structure of the adjoining noun:

Fig. XIV.

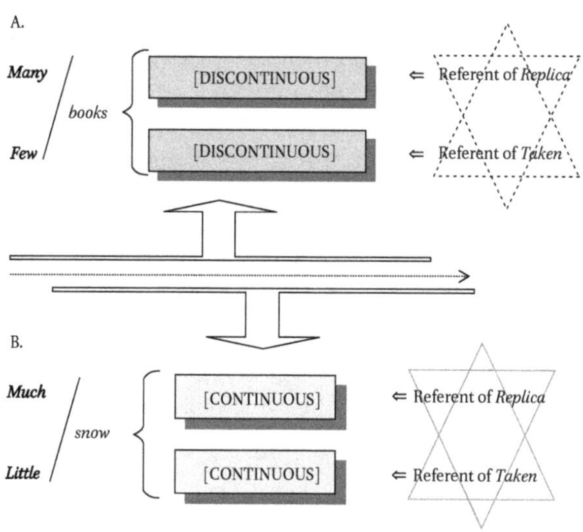

Fig. XIV (A)–(B) renders a perfect explanation of the genuine cause of what Michael Bennett characterized as "the remarkable parallelism between the plural quantifiers and the quantifiers for mass nouns."[12] This notable parallelism in the inventory of English quantifiers, namely the existence of the quantifier pairings such as *many—much, few—little, a few—a little, fewer—less,* the pursuit into the nature of which has evoked much of recent philosophical debate on mass terms,[13] we may now present in the way as shown in Fig. XV below, where $\frac{DC}{C}$, $\frac{DC}{DC}$, and $\frac{C}{C}$ (C = continuity, DC = discontinuity) are the respective markers standing for the basic relational characteristics of Thirdness:[14]

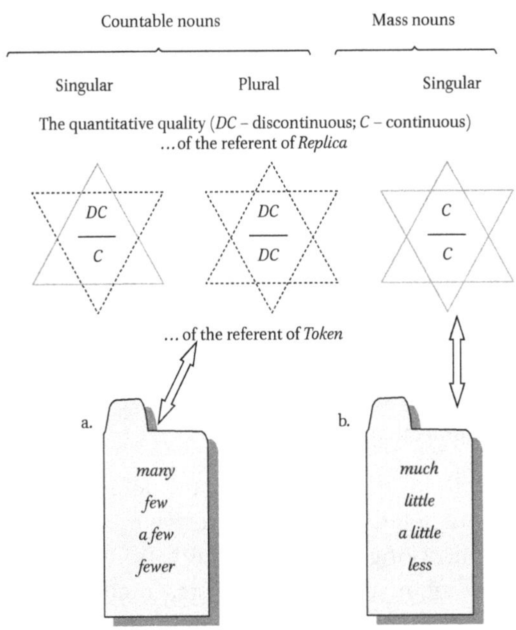

Fig. XV.

It is easy to see that the related quantifier items from (a) and (b) above appear to share the same assessment of the degree of quantity[15] as well as the same (positive) assessment of the homogeneity, although the type of

---

[12] Michael Bennett, *Mass Nouns and Mass Terms in Montague Grammar*, p. 264.
[13] See Laycock 1975, 1998; Chellas 1979; Pelletier 1979; Bunt 1985; Zimmerman 1995.
[14] See also diagrams in Fig. XII (A), (B), (C), respectively.
[15] I.e., the one rendered uniquely by their translation equivalent in Russian.

homogeneity that they refer to differs as to whether the homogeneity in question is that of discontinuity or continuity.

4.2. *Vendler's Query: 'Each' and 'Every', 'Any' and 'All'*

A related problem arises in the attempts at translating English quantifiers into the formalism of Frege-Russellian logic. The notorious insufficiency of quantification theory as a representational device for natural language quantifiers was first touched upon in *The Principles of Mathematics* (1903) by Russell, who was aware that the differences between lexical quantifier items or else particles of quantification, such as *all, every, any, some*, is a feature of natural languages which has no obvious counterpart in formal logic. Sixty-four years later this issue was pressed yet further by Vendler, who demonstrated that the method of lumping such particles of quantification as *each, every, any* and *all* together and treating them as merely stylistic variants of the universal quantifier ∀ tends to suppress certain logically important aspects (manifested by their lexical differences) that enter into the common understanding of these words. Thus, as Vendler remarks,

> However this may be, our results are sufficient to show that a simple application of the theory of quantification may fall short of capturing all the logically relevant features involved in the vernacular use of the particles of quantification. Some such features can be found by contrasting *all* with *each* and *every*, but the most important points missed by the theory are the ones that emerge in connection with *any*.[16]

Before turning directly to the most important points arising in connection with *any* in more detail, consider the logically relevant features that can be found by contrasting *all*, on the one hand, with *each* and *every*, on the other. According to Vendler, the representation of all of them in terms of the same logical structure tends to obscure issues concerning the type of reference that turn out to be logically relevant. Thus, comparing (2.1)–(2.4):

(2.1) All those blocks are yellow,
(2.2) All those blocks are similar,
(2.3) All those blocks fit together,
(2.4) The number of all those blocks is 17

[16] Z. Vendler, *Each and Every, Any and All*, p. 96.

with their respective counterparts where *all* is replaced by *each* or *every*, he shows that (2.1) is true if and only if the proposition

(2.1') Each (every one) of those blocks is yellow

is true. So in this case, at least in so far as truth-values are concerned, no difference appears between the functions of these particles.

This, however, is obviously not the case in regard to (2.2)–(2.2'):

(2.2') Each (every one) of those blocks is similar (to every other).

Since, as Vendler notes, it is quite possible that each block be similar to every other without all of them being similar, (2.2) entails (2.2') but not *vice versa*. It is also possible that each block fits every other without all of them fitting together. Therefore, the same holds also in the case of (2.3)–(2.3'):

(2.3') Each (every one) of those blocks fits every other.

Proposition (2.4) brings out the difference in the most extreme form. The counterpart sentence

(2.4') The number of each (every one) of those blocks is 17

will not make sense unless an entirely different interpretation of *number of* is invoked—being marked, say, with the numeral *17*. In which case, of course, there is no logical relation between (2.4) and (2.4') whatever. According to Vendler, these examples show that

> The relation of similarity (with the given interpretation) and of fitting together can apply to the whole set in a *collective* sense, or to subsets (couples) of the whole group in a *distributive* sense; and the expressions *are similar* or *fit together* do not indicate, by themselves, in which of these senses they be predicated. It is, therefore, up to the quantifier particles alone to decide the issue.[17]

As Vendler concludes,

> Since, however, the collective sense may fail to imply the distributive sense and vice versa, that is to say, one respective proposition may be true and

[17] *Ibid.*, p. 74.

> the other false, such a difference in truth-values clearly indicates the difference in the meaning of these particles. Similarly, in the last case, the phrase *number of* requires an entirely different interpretation according to whether collective or distributive reference is indicated by the quantifying particle. We can safely conclude then that, at least with respect to a given group of individuals, the reference appropriate to *all* is collective, and the reference appropriate to *each* or *every* is distributive.[18]

Moreover, Vendler goes on to point out that his "... considerations in this section not only confirm the basic difference between the collective *all* and the distributive *every* and *each*, but they suggest a divergence in the respective functions of the last two particles as well."[19] Vendler illustrates this by his examples in (2.5)–(2.6):

(2.5) Every deputy rose as the king entered the House
(2.6) Each deputy rose as his name was called

According to Vendler, *every* in (2.5) stresses completeness or, rather, exhaustiveness (like *one man* they rose); whereas *each* in (2.6) directs one's attention to the individuals as they appear, in some succession or other, one by one (each deputy rose as *his* name was called).[20] This time, however, Vendler appears to go on talking about the same distinction to which he refers in only slightly different terms as otherwise he would be compelled into talking about *every* indicating once collective and another time distributive reference. Besides, in Vendler's own terms, *all* does not properly fit the opposition between collective reference and distributive reference either, since there can be none such in the case of uncountable, or "mass" nouns with which *all* occurs in (2.7)–(2.8):

(2.7) All the information we obtained was worthless.
(2.8) All petroleum is organic in origin.

The paradoxical ambiguity that arises between the collective and the distributive readings of *every* can evidently not be resolved in terms of Vendler's distinction alone. Neither can the striking asymmetry of *all* be explained. All of this can be straightforwardly accounted for in terms of synthetic *a priori* structures falling under the category of Thirdness, however, by appealing to qualitative differences between the characteristics

---

[18] *Ibid.*
[19] *Ibid.*, pp. 78–79.
[20] *Ibid.*, p. 78.

of the relation of Thirdness that these quantifiers indicate. In the case of *all*, the latter can be described as the identity of either of the basic characteristics, quantitative and qualitative, of the quantity assigned to both objects of reference. By contrast, in the case of both *every* and *each* it can be rendered as the identity of its *quantitative* despite the non-identity of its *qualitative* characteristics. Thus, all things being equal, the difference in question boils down to a difference in one feature only, that between the qualitative (b) homogeneity and (a) non-homogeneity of Thirdness, i.e., the *simultaneous discontinuity* or else *continuity* of both referents *vs.* the *discontinuity* of the referent of the Replica as opposed to the *continuity* of the referent of the Token:

Fig. XVI.

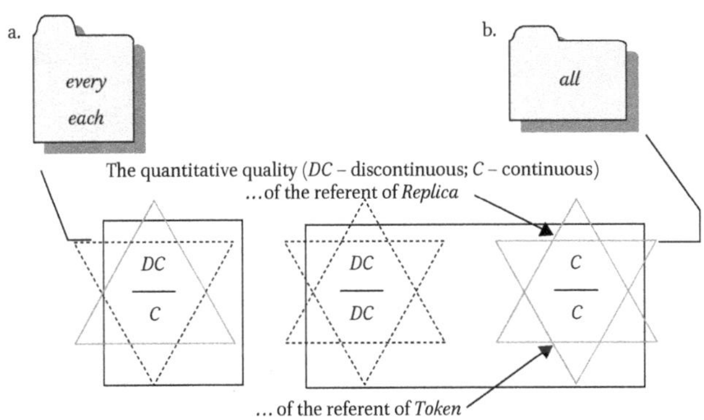

It would then be immediately evident that the nature of the distinction between *every* and *each* itself is likewise determined by the same kind of difference, only this time no issue of the presence or else lack of qualitative homogeneity can intelligibly be clarified without appealing to yet *another instance of Thirdness*. The relevant Thirdness now turns out to be brought into play by (b) the verb phrase which is adjoined to (a) the noun phrase where *every* or *each* occur as noun determiners (Fig. XVII):

Fig. XVII.

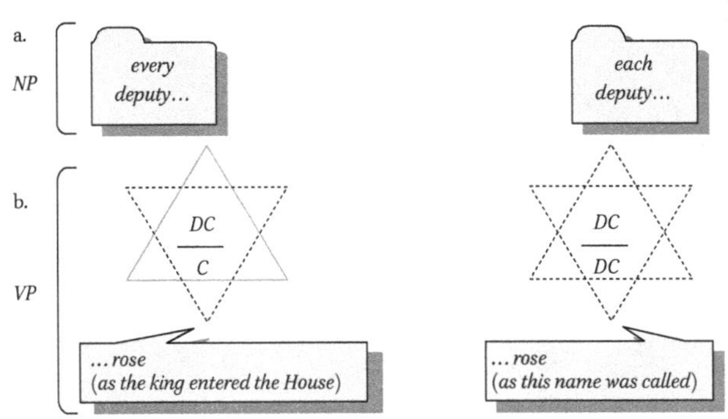

As is shown in yet more detail in Fig. XVIII below, the *discontinuity* of the referent of the Replica, only this time of the *verb 'to rise'*,[21] is now contrasted with the *continuity* of the referent of its Token when this verb occurs with *every deputy*,[22] as opposed to the case of its occurrence with *each deputy* where we may observe the like *discontinuity* in both its referents:[23]

Fig. XVIII.

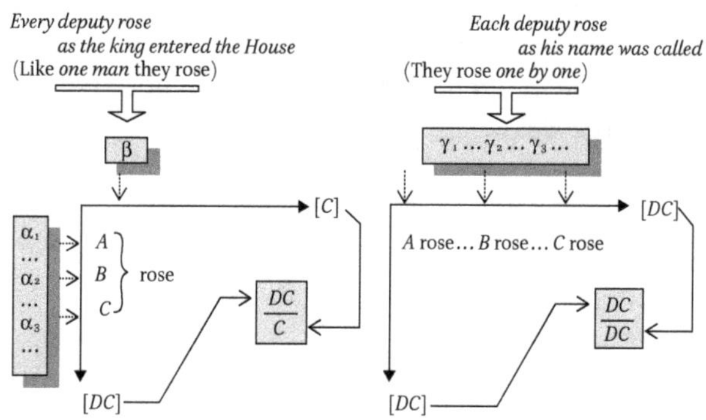

If this analysis is correct, its results make it possible for us to turn directly to the most important points raised by Vendler—those that emerge in connection with *any*. By now, in the light of the foregoing analysis, the points made by Vendler in connection with *any* will appear to be more significant than they seemed to be before. Moreover, in view of this our theory seems for the moment to be in a uniquely favourable position not only "to open up a new line of attack on the problem of lawlike propositions"[24] (and show that, quite contrary to Frege, analytic *a priori* is, at bottom, itself synthetic *a priori*), but also to develop as the study of highly abstract principles and structures that turn out to determine the so-called partitive vs. non-partitive use of quantifiers. The latter issue happens to be the most notorious moot point in the history of quantifier analysis ever since Jespersen—the moot point regarding which logicians have the least grounds of all to pride themselves on having won through to the true explanations of expressions that had previously defied

---

[21] ...visualized as a multiplicity of projection points of *A*, *B*, *C*, ...*k* onto the vertical axis ($\alpha_1 \dots \alpha_2 \dots \alpha_3$).

[22] ...visualized as a single indivisible projection point of *A*, *B*, *C* onto the horizontal axis (β).

[23] Cf. multiple projection points of *A*, *B*, *C* onto the horizontal axis ($\gamma_1 \dots \gamma_2 \dots \gamma_3$).

[24] Z. Vendler, *Each and Every, Any and All*, p. 96.

analysis.[25] The reason for this failure will be seen in the fact that in the light of the theory of formal systems, the quantified statements in question and the true-but-unprovable statements of Godel's Incompleteness Theorem are isomorphically equivalent.

In the wider context of Peirce's categories, the essence of the most important points that Vendler raises as missed by the theory that emerge in connection with *any* looms large, quite naturally, in the light of Vendler's own observations concerning the use of *any*. If we were to summarise the points made by Vendler in connection with the different uses of *any*, namely its use in questions, future tense and, generally, in non-factive contexts, we can easily see that the most general characteristics of *any* is that the referent of its adjunct cannot be conceived as a certain physical object with specific space-time location. The very fact that *any* may only occur in constructions lacking definite reference and existential import can be seen as a clear indication that the Token-referent of the construction in question gets devoid of its Token-like properties and acquires the essential characteristics of the referent of Replica.[26] This is even more vivid in the light of Vendler's conclusion concerning the use of *any* and (partly) of the use of *all*, in which he appeals to the difference between laws and statements of fact:

> Thus the nonreferential *all*-proposition, in much the same way as the nonreferential *any*-proposition, cannot be found true as a result of enumerative induction. Such propositions always remain open, whereas statements of evidence, statements of fact, are necessarily closed. Laws are not statements of fact and statements of fact are not laws.[27]

Now, since this distinction drawn in the cited passage by Vendler and Peirce's distinction between *sinsign* (Token) and *legisign* (Type) are absolutely identical, we may safely conclude that the role of *any* amounts to no more and no less than the duplication of the Replica of the Sign in its Token or, to put it alternatively, to representing specific *Token*-referents the way *as if* they were *Type*-referents.

In view of Vendler's distinction between the non-referential use of *any* and the use of *every* in statements of evidence, statements of fact, the

---

[25] O. Jespersen, *Analytic Syntax*, p. 118; G. Carden, *English Quantifiers: Logical Structure and Linguistic Variation*, p. 1; R. Hogg, *Quantifiers and Possessives*, pp. 227–32; D. A. Lee, *Quantifiers and Identity in Relativisation*, pp. 1–19; T. Thrane, *Referential-Semantic Analysis: Aspects of a Theory of Linguistic Reference*, p. 127.

[26] Z. Vendler, *Each and Every, Any and All*, p. 91.

[27] *Ibid.*, pp. 93–94.

corresponding distinction between the narrow-scope use of *every* and the wide-scope use of *any*, made by Quine, comes as a mere confirmation of this very fact, only appearing in the disguise of talk about "scope." Thus, in Quine's own terms:

> ... the difference in English usage between 'any' and 'every' may have struck many of us as unsystematic and even mysterious. Scope of the universal quantifier is the key to it. Where distinction is needed between broader and narrower scope, as between '$(x)(Fx \supset p)$' and '$(x)Fx \supset p$', the English speaker's unconscious understanding is that 'any' calls for the broader scope and 'every' for the narrower. This rule works not only in connection with the conditional but also elsewhere, notably in connection with negation. Thus take the universe of discourse as consisting of all poems. 'I do not know any poem', then, and 'I do not know every poem', call respectively for the broader and the narrower scope:
>
> $\forall x \sim$ (I know x), $\sim \forall x$ (I know x)[28]

If we look at Quine's explanations in the light of the foregoing discussion, it will be readily noticed that Quine's use of negation falling within the quantifier scope (2.9) serves no other end than just epitomize the universal unattainability of the *x*'s in question (as contrasted with the use of the quantifier appearing, in the reverse order, within the scope of negation). In Peirce's terms, their universal unattainability and, thus, non-facticity amounts to none other than just defining, in (2.9), the status of the referent of 'poem' as that of *legisign*:

(2.9) $\forall x \sim$ (I know x) 'I do not know any poem'

This is naturally opposed to saying, in (2.9'), that the *x*'s in question are such that, *as a matter of fact*, not all of them are familiar to me:

(2.9') $\sim \forall x$ (I know x) 'I do not know every poem'.

All this goes to show that the presence of *any* comes to signal that the relation of Thirdness in the co-occurring nominal item is one between the two Replicas of the same Type. This is not to say, however, that the relation is that of complete identity. Not only in the sense that the Token, even though it acquires Type-essential characteristics, yet functions as a Token, but also due to the fact that the Type's duplicate that acts as a

[28] W.V. Quine, *Methods of Logic*, p. 120.

Token can also differ from its original in some of its characteristics. This is best shown when the co-occurring item is a singular noun, as in *any doctor* in Vendler's example, *Any doctor will tell you that Stopsneeze helps*:

Fig. XIX.

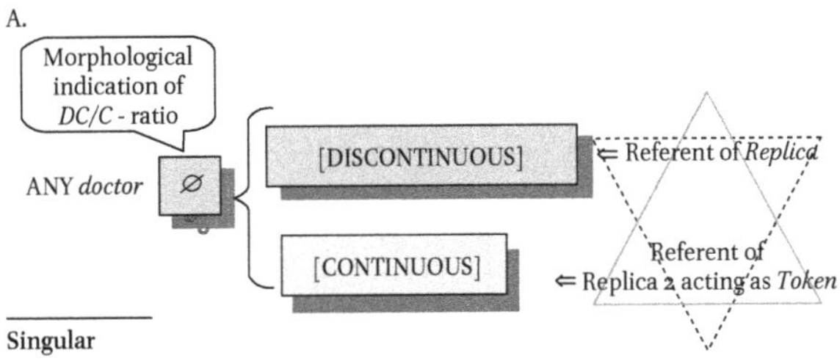

The same, of course, will also be true of the so-called partitive use of quantifiers, as in the corresponding example of Vendler's:

> *Any one of the doctors (that you ask) will tell you that Stopsneeze helps*

This is not so, however, in the case of *any doctors*, meaning, essentially, the same as

> *Any group (no matter how large) of doctors will tell you that Stopsneeze helps*

where the qualitative characteristics (though not purely quantitative ones) of the quantity of both referents happen to be strictly identical:

Fig. XX.

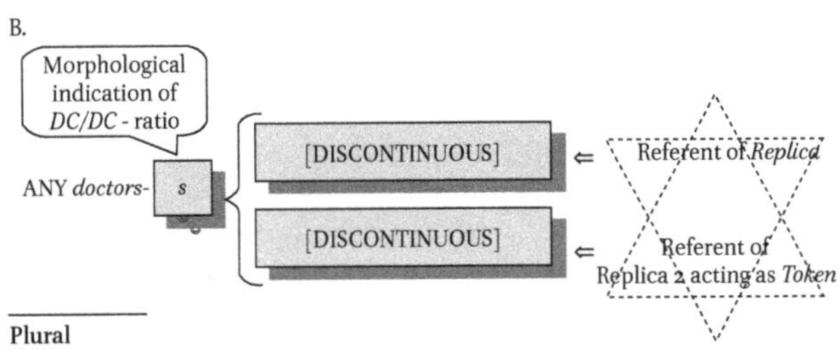

Here we have a distinction between the different uses of *any* according to whether the co-occurring noun is a) a singular noun, or b) a plural noun or a mass-noun:

Fig. XXI.

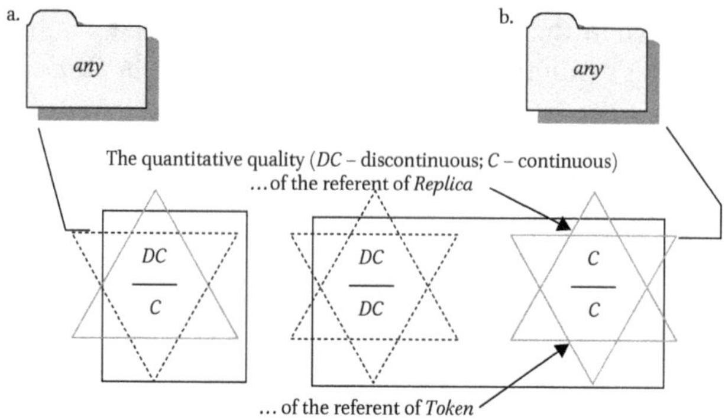

4.3. *Further Keys to Addressing Quantification: The Analysis of Non-Partitive vs. Partitive Use of Quantifiers*

There appears to be no important difference between the partitive and non-partitive use of *any*, insofar as the distinctions between the $\frac{DC}{C}$-, $\frac{DC}{DC}$-, or $\frac{C}{C}$- ratios exemplified in Fig. XXI might be concerned. On the face of it, there linger further distinctions lurching around, evolving this time strictly *within* the partitive/non-partitive dichotomy itself, which have to do, *prima facie*, with the referential devices mentioned by Vendler. The latter are deictic elements that mark the "existential" use of *all*, as opposed to its so-called "non-existential" use:

> We have to say... that while *each* and *every* always connote existence, *all*, by itself, does not. It may occur, however,... in propositions that do have existential import due to some other referential device which may be joined to *all* within the same noun phrase (definite article, demonstrative or possessive pronoun, etc.) This possibility is not available with *any*: we do not have *any the*..., *any my*..., etc. We have to say, for example, *any one of the*... – that is, we put the definite article into a separate noun phrase which then will carry existential import.[29]

In Vendler's examples, the use of this referential device appears indispensable when *all* is used in a factive (existential) context:

(3.1) All *the* messages you sent were intercepted

[29] Z. Vendler, *Each and Every, Any and All*, p. 91.

By contrast, the use of the definite article in the same noun group is prohibited when *all* occurs in a non-factive (non-existential) context, in which *all* is interchangeable with *any*:

(3.2) *All* messages you might have sent would have been intercepted
(3.3) *Any* message you might have sent would have been intercepted

Since *every*, unlike *all*, is admissible only in a factive context:

(3.4) Every message *you sent were intercepted*

but is not acceptable in a non-factive context:

(3.5) * Every message *you might have sent would have been intercepted*

the use of the definite article in (3.4) in the role of a disambiguation device, as opposed to (3.1), is superfluous. Accordingly, the Type/Token ratio, $\frac{DC}{C}$, is the same in both (3.3) and (3.4), regardless of whether the context is factive, as in (3.4), or non-factive, as in (3.3). The only difference between the "non-referential" reading in (3.3) and the "referential" one in (3.4) is that the continuity of the Token in (3.3) inheres in the *Type*-like referent. This is clearly seen on considering the behaviour of the 'lazy' pronoun *one* that occurs in the partitive construction with *any* as well as *every*, thus sharing the same ambiguity as *all* with respect to the "facticity" or else "non-facticity" of the referent of Token:

(3.6) Every *one* of the messages *you sent were intercepted*
(3.7) Any *one* of the messages *you might have sent would have been intercepted*

Contrary to Vendler, however, the occurrence of the definite article in a separate noun phrase, the messages, as it happens in (3.7):

(3.7) Any one of ***the*** *messages* you might have sent would have been intercepted

does not mark that this noun phrase carries existential import. For the definite article in appearing in the structure of a separate noun phrase in:

(3.8) (any one)$_{NP}$ of (the messages)$_{NP}$. you might have sent

does not render the meaning of (3.8) any different from that of (3.9):

(3.9) (any message)$_{NP}$ you might have sent.

It will be evident that both (3.3) and (3.7):

(3.3) *Any message* you might have sent would have been intercepted
(3.7) *Any one of the messages* you might have sent would have been intercepted

equally express unreal condition, quite regardless of the occurrence of the prepositional *of*-phrase that contains the definite article. Thus we can see that the use of the definite article in (3.7) marks the existential import to no more extent than, e.g., the use of the definite article in Russell's definite description in:

(3.10) The present king of France is wise

Moreover, the use of the definite article happens to be ambiguous between referential and non-referential readings in exactly the same way as yet another item on the list of what Vendler speaks of as *referential* devices—the possessive pronoun 'his', cf. (3.6–3.7) and (3.11)–(3.12):

(3.6) Every *one* of **the** messages *you sent were intercepted*
(3.7) Any *one* of **the** messages *you might have sent would have been intercepted*
(3.11) Bill is *kissing* **his** future wife
(3.12) Bill is *thinking* of **his** future wife

The reason for this is that a definite description is not an independent unit in the statement at all. For, on Russell's analysis, the statement in

(3.10) The present king of France is wise

is shown to be a complex conjunction of statements: (1) "There is a present king of France"; (2) "There is at most one present king of France"; and (3) "If anyone is a present king of France, he is wise." But, more importantly, each of the three components makes a general statement that, according to Russell, is not about anything or anyone in particular. Thus, according to Russell's theory of definite descriptions, there is no phrase in the ultimate analysis of (3.10) that would be equivalent to 'the present king of France'. This shows that, despite the presence of the definite article, the phrase is not an expression that refers, like a proper name, to something as the thing that the whole statement talks about. Analogously, "his future wife" in (3.11)–(3.12) does not turn out to be an independent unit of analysis, either. In the latter case, the "referential" or "non-referential" interpretation of "his future wife" depends on whether this noun phrase appears in a so-called extensional context (of the verb *to kiss*), or else in an intensional context (of the verb *to think*), in which case we are told that what *Bill's future wife* designates in one of the readings of (3.12) is an individual concept, not an individual. Accordingly, the analysis will show that in the "quantified" sentences:

(3.6) Every one of *the* messages you sent were intercepted
(3.7) Any one of *the* messages you might have sent would have been intercepted

the occurrence of the definite article, in the same way, has nothing to do with giving evidence to the effect that the noun phrase carries existential import. On the one hand, the definite article in the prepositional *of*-phrases above will be seen to provide no more evidence to that effect than it does in Russell's definite description, "the present king of France." Nor the known contextual considerations, on the other hand, taking into account the context of the verb as in the case of *Bill is thinking of his future wife* will do either. For, in the case of the verbs *to send* or *to intercept* that occur in (3.6)–(3.7), reference to the "intensional" (as opposed to the "extensional") reading of these verbs—which amounts to believing that an individual concept can be not only "thought of" but also "sent" as well as "intercepted"—can hardly make any sense at all.

All this goes to show that what can count as an intelligible explanation can only be arrived at by giving more scrutiny to the roles of the structural positions within the prepositional *of*-phrase that provides the general pattern for both (3.6) and (3.7) alike. In both (3.6) and (3.7), the prepositional *of*-phrase in question contains definite descriptions, "the messages you might have sent" and "the messages you sent," in the structural position to the right of the preposition *of*. According to *Encyclopaedia Britannica*, definite descriptions present themselves as expressions (noun phrases) that, despite the presence of the definite article, are not about anything or anyone in particular.[30] However, what we may only have by way of depicting definite descriptions in this manner is a rather vague and, in the first place, negative characterization of definite descriptions as such. This characterization can nonetheless be made more explicit and more specific if we define them in Peirce's terms as noun phrases that, despite the presence of the definite article, are *not* about the *Token* referent of the expression. As a result, we can proceed in our dealing with the definite descriptions at hand in a more economical way, avoiding Russell's elaborate technique of defining definite descriptions as expressions that are analysable into general statements. And, by doing so, we are turning half way towards the positive characterization of the definite descriptions that occur in the rightmost position of a quantified *of*-phrase. Namely, we are now in a position to deduce, by *modus tollendo ponens*, that these expressions are such that they are, therefore, *about* the referent of *Type*. More specifically, because these expressions can be shown to acquire their specific referential properties owing solely to their occurrence as elements of a specific context, we can now characterize them as expressions that designate a *Type* referent of the *quantifier matrix*, that is, of the *whole* nominal phrase that immediately follows the quantifier. In other words, the structural role of the rightmost position in the prepositional *of*-phrase under consideration amounts to signalling that the Token referent of a single element of the quantifier matrix, i.e. that of the NP that occurs to the right of the preposition, now takes over, and thus is upgraded to, the role of the respective Type referent of the quantifier matrix as a whole. In this way, the quantifier matrix in question, no matter how expanded syntactically, appears to stand semantically as a singular designation, in the same manner as it is being done by a single noun when it occurs in simple juxtaposition with the quantifier (when the quantifier complement is not expanded into a nominal group with more than one structural element):[31]

(3.13) Quant + Complement

(i) Quant + N_{COMP} — Simple (Synthetic) Quantifier Complement

(ii) Quant + $(one + of + Det + N)_{COMP}$ — Complex (Analytic) Quantifier Complement

---

[30] Vide "Analytic philosophy" *Encyclopædia Britannica.*

[31] Cf. A. Molczanow, *Set Reference Relationships and the Phrasal Syntax of Quantifiers in English*, pp. 88–89.

Further, the aforementioned phenomenon of "reassignment" of the referent of Type to the role of the referent of Token, in this setting, presents itself as fully symmetrical: what we are having now is the same kind of reassignment of referential roles as shown previously that merely proceeds in the opposite direction (the arrows show the direction of reassignment).

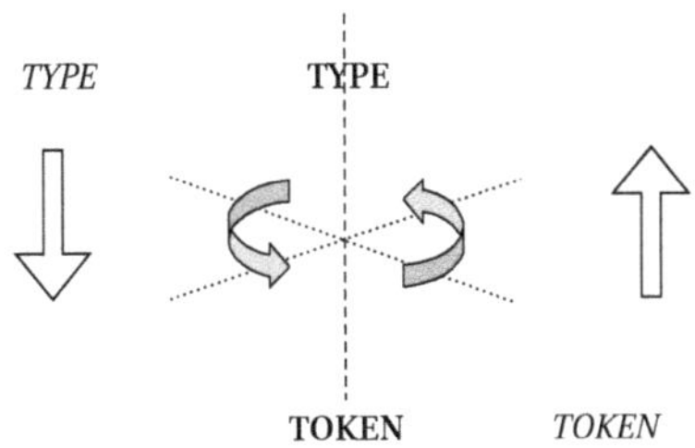

The general conclusion, thus, is that the ultimate analysis of *any one of the messages*, in its very roots, appears to be the same as that of *any message*. All this gives grounds to contend that the analysis of the quantifier matrix in (3.7), *one of the messages*, should proceed in exactly the same way as the analysis of the quantifier matrix in (3.3), *message*. Apparent differences dissolve, on enquiry, to alternative surface realizations of the sole underlying synthetic *a priori* structure associated with the relation of Thirdness. The Thirdness in question, which in (3.3) appears as a relation between the different types of referents of the same noun:

Fig. XXII.

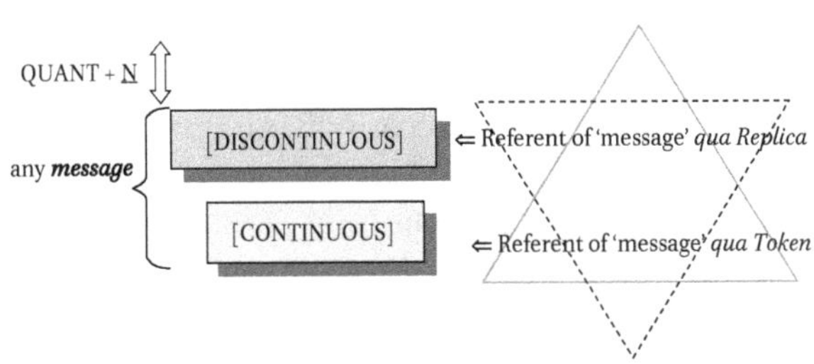

now presents itself, in (3.7), as the same relation,—even though it occurs to be the one between the referents of structurally *separate* items that appear in *different* structural positions:

Fig. XXIII.

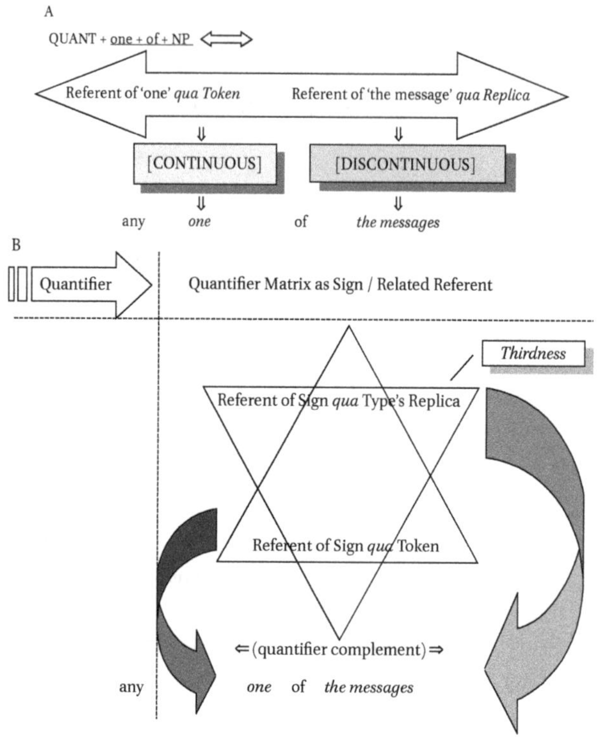

In Fig. XXIII, the separate items that occur in the structural positions of the expanded quantifier matrix to the left and to the right of the preposition *of* can now be seen to assume, respectively, the semantic roles of Token and Type that in the syntactically trivial case of (3.13i), as shown in Fig. XXII, appear as roles assumed by the same syntactic item. In this way, the difference between the two structural realizations of the quantifier~matrix configuration in (3.13), known as the "partitive" (3.13ii) and the "non-partitive" (3.13i) uses of the quantifier, can be explained in terms of a binary opposition between the two types of 'alignment' of the Type/Token relation, i.e., 'vertical' (Fig. XXII) *vs.* 'horizontal' (Fig. XXIII). In other words, what we can now observe is a sort of "paradigmatic" interconnection of Type and Token in the case when the quantifier complement is syntactically primitive (as in 'any *message*'),[32] as opposed to the "syntagmatic" interconnec-

[32] ... when Type and Token occur as the respective referents of the same structural item.

tion in the case of the syntactically expanded quantifier complement (as in 'any one of the messages'). In the latter case, the Token referent of the rightmost NP acquires a dual role. On the *intrasegmental* level, its role is that of the Token referent of 'messages' *per se*. On the *segmental* level, however, on which the quantifier complement presents itself—regardless of whether it is simple or expanded—as a single nominative unit, the role of the referent in question shifts to the role of the Type referent of 'one of the messages' taken as a whole.

4.4. *Earlier Proposals for Quantifiers*

The treatment of quantifiers in their partitive and non-partitive use has always been open to debate—albeit on various preconceptions about the nature of linguistic signs, grammatical meaning, and explanations of meaning—ever since Jespersen first considered them in these two syntactic environments (later to become known as diagnostic environments of quantifiers) in the context of his theory of rank.[33] In his *Analytic Syntax*, Jespersen observes four apparently diverse ways of combining a quantifier (q) with a quantified item (Q):

(a) qQ, q as adjective to Q: many girls, five girls.
(b) qQ, q as a substantive: G. ein glas wein, Dan. et glas vin.
(c) qQ, different from (b) by being in the genitive: L. poculum vini; Ru. stakan vina.
(d) qpQ, a glass of wine, F. un verre de vin.

According to Jespersen, both (c) and (d) originally started from a partitive idea (part of a definite quantity) as in L. *pars militum*, F. *un grand nombre de nos amis*, but they were by an easy extension applied to an indefinite quantity (genetivus generis) as in the examples given above. In Jespersen's view, in both cases the extended use was promoted by the general disinclination to an immediate collocation of two substantives in the same case. Thus, while we have a partitive idea in *many of us, three of us*, there is no partitive idea in *all of us, the three of us*, to which the symbol for apposition with *of*, was applied above.[34]

[33] ... more on the partitive and non-partitive use of quantifiers as presenting the so-called diagnostic environments of quantifiers, see Thrane, *Referential-Semantic Analysis: Aspects of a Theory of Linguistic Reference*, p. 127.
[34] O. Jespersen, *Analytic Syntax*, pp. 117–18.

Notice that the q above is treated as an adjective in (a) '*many* girls', whereas in (b) '*ein glas* wein', as a substantive. This poses a basic problem for the treatment of quantifiers in Jespersen's theory of rank, since substantives are classified as items of primary rank, while adjectives, as their adjuncts, thus, as secondary items, or secondaries. So, while in (a) q is treated as a secondary, Q as a primary, in (c) and (d) it is inversely q that is treated as the primary, and Q is expressed in two ways that are generally used for secondaries, whereas the collocation in (b) shows in itself nothing about the rank. On the face of it, anyway, Jespersen reaches an arbitrary decision to treat a quantifier in both non-partitive and partitive uses, such as *many* occurring either in *many girls* or *many of the girls*, in a unified way, namely, as a secondary. This, however, creates immediate tension in Jespersen's theory, since elsewhere he himself manifestly indicates that it is not the case with quantifiers that they—like ordinary adjuncts—define a primary, and points out that in the analysis of quantifiers we meet with a *special sort of relations, different from those in junctions*.[35] Thus, despite being quite aware of this special status and yet undisclosed nature of quantifiers, Jespersen seems to be utterly misguided by what he views as a profound philosophical importance of his theory of rank—to the extent which leads him away from seeing all this as an evident sign of its inadequacy. Thus, according to Jespersen, the tension that—as he himself states—"goes to show the difficulty of applying the theory of ranks outside of qualifiers" turns out eventually to be attributed, by mere prejudice, to the assumed logical inadequacy of linguistic form. No wonder, however, since this was a preconception that was widely held at the time of Jespersen's writing. According to Jespersen's own explanation:

> After a good deal of hesitation I have finally adopted the plan of *everywhere taking the quantifier as secondary and the quantified as primary*, no matter how expressed. This is nothing but a consistent carrying out of the general principle of disregarding form and penetrating behind it to the notional (or, if you like, the logical) kernel of the matter.[36]

Thus, the analysis of quantifiers provided within Jespersen's theory of rank presents yet another instructive object lesson, which hardly shows that the lexical and the logical can be for all purposes amalgamated in some unitary idea of the logical structure or skeleton of a sentence that

---

[35] *Ibid.*

[36] *Ibid.*, p. 119. No wonder this view is still persistent in modern scholarship in the guise of the theory of "quantifiers as higher predicates."

is rendered, or construed, exclusively in terms of relationships *between ideas or concepts*, without due consideration of the relation of reference to objects that is the main rationale of Kant's transcendental logic.

In the years that followed the publication of *Analytic Syntax*, the problem of quantifiers was still in the centre of much discussion, though in a slightly reshuffled form. The subsequent shift, in the years to come, to the agenda of structural linguistics and the rise of generative-transformational grammar led to the rejection of Jespersen's view on quantifiers as notional words. In the structuralist framework after Fries, the status of quantifiers was changed to that of purely structural signals and thus strictly grammatical form-words, so Jespersen's view on the relation between the quantifier and the quantified item as an outward manifestation of an underlying *relation between ideas* of different rank, one of which was the *idea of quantity*, was dropped. The relevant question was then reformulated as one regarding the nature of their meaning as structural signals in the constituent structure of the quantified noun phrases.

In the earliest transformational approach to quantifiers, developed by Chomsky and elaborated by Barbara Hall, quantifiers were analysed as part of the determiner sequence followed by a head noun, so the syntactical analysis of the prepositional phrases with quantifiers looked as shown in (4.1).[37]

(4.1) *some of the men*

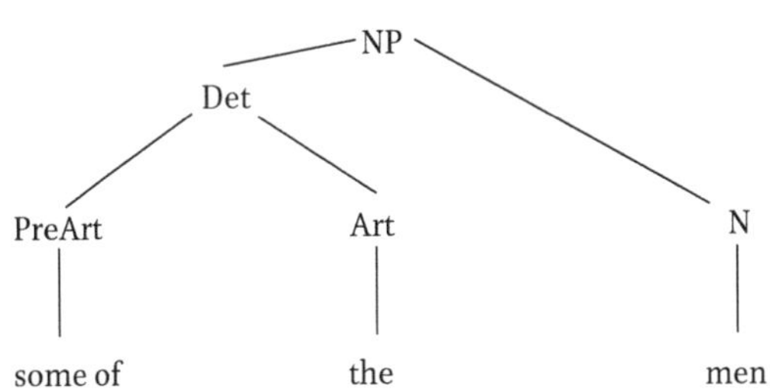

The obvious advantage of this approach was in that it abandoned the earlier counterintuitive analysis of quantifiers in terms of various parts of speech. In traditional grammar, the most salient feature of the treatment of quantifiers was a good deal of confusion as to their status in terms of traditional parts of speech, confusion that persists in lexicographical

[37] Barbara Hall, *All About Predeterminers*, unpublished M.I.T. paper.

practice even today. Traditionally, quantifiers were often classed either with adjectives or with (indefinite) pronouns, according to their occurrence x in either of the two environments in the diagnostic frame in (4.2):[38]

(4.2) (a) x Noun

$$\text{(b) } x \quad \text{of} \begin{Bmatrix} \text{these} \\ \text{this} \\ \emptyset \end{Bmatrix} \text{Noun}$$

The problem with this approach was that the same quantifier had to be treated in terms of two different parts of speech, depending on its occurrence in either (4.2)(a) or (b). In this case, the quantifier *many* (adj) that occurs in the position of x in (4.2)(a), as in *many men*, should be regarded as a distinct lexical item, different from that of *many* (prn) in the position of x in (4.2)(b), as in *many of the men*, which is an evidently counterintuitive solution. To avoid this cumbersome situation, the usual practice was to refer to certain quantifiers, like *some*, as indefinite pronouns, while other quantifiers, like *many*, were regarded as adjectives. Nevertheless, this solution was not satisfactory either.

Despite the manifest advantage of the structuralist approach employed in Hall's analysis in rejecting the dubious classification of quantifiers in terms of various parts of speech and categorizing them, instead, as noun determiners regardless of their position in either (4.2)(a) or (b), Hall's analysis, however, had serious drawbacks. First, the analysis was inconsistent with the received treatment of quantifiers as a subclass of determiners, since in the determiner sequence before the head noun it obligatorily put all quantifiers in the position of a predeterminer, which went contrary to the established classification of quantifiers into predeterminers, central determiners and postdeterminers.[39] Thus, in (4.1), *some* had to be treated as a predeterminer, though in the structuralist classification it belonged to the subclass of central determiners, of which no member could occur to the left of the article by definition.[40] Besides, the idea of analysing *some of the* as a determiner sequence ran contrary to facts of noun-verb number agreement. They clearly indicated that the word that followed the deter-

[38] The diagnostic frame in (4.2) also provides the basis for the structural definition of quantifiers which are defined as items "that may occur in place of x in both [4.2](a) and (b)" (T. Thrane, *Referential-Semantic* Analysis, p. 127).

[39] R. Jackendoff, *Quantifiers in English*, p. 430.

[40] See e.g. Leech and Svartvik, 1975, section 550.

miner sequence was by no means the head word of the group (which also presented a major problem with Jespersen's analysis of quantifiers). Moreover, the proposal of analyzing *some of the* and *men* as the immediate constituents in *some of the men* was inconsistent with linguistic evidence that indicated the possibility of preposing the whole of the prepositional phrase to the quantifier. Nevertheless, the requirement on Hall's proposal was that *of the men* need not be a constituent, „thus immediately losing the generalization that it is a plain ordinary prepositional phrase."[41] Since putting the preposition *of* in the determiner sequence, among other things, violated the requirement that, for the sake of maximal generality, *of* should be rather viewed as part of a plain ordinary prepositional phrase, as in sequences like *the scene of the movie*, where *of the movie* is a constituent:

(4.3) *the scene of the movie*

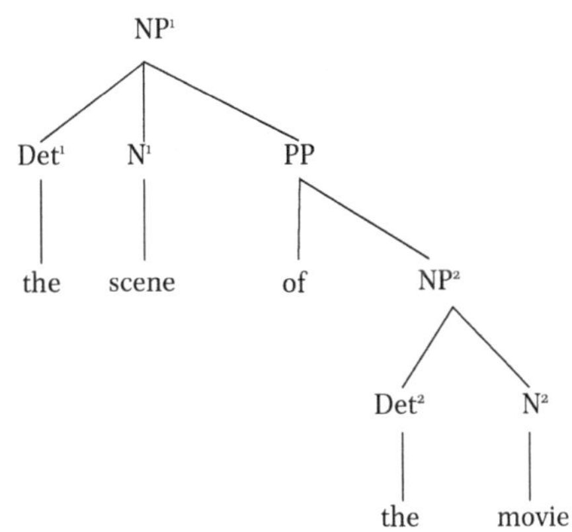

while *of the men* in (4.1) is obviously not, this type of analysis was, however, dropped from grammars. Instead, two alternative analyses were developed in the late 60's, both satisfying the requirement that *of the men* in (4.1) should be treated as an ordinary prepositional phrase, on a par with *of the movie* in (4.3). The first, put forward in a number of studies by Dean, Postal, Langacker, suggested derivation of phrases like *some of the men* from the underlying structures as shown in (4.4):

[41] R. Jackendoff, *Quantifiers in English*, p. 430.

(4.4) *some of the men*

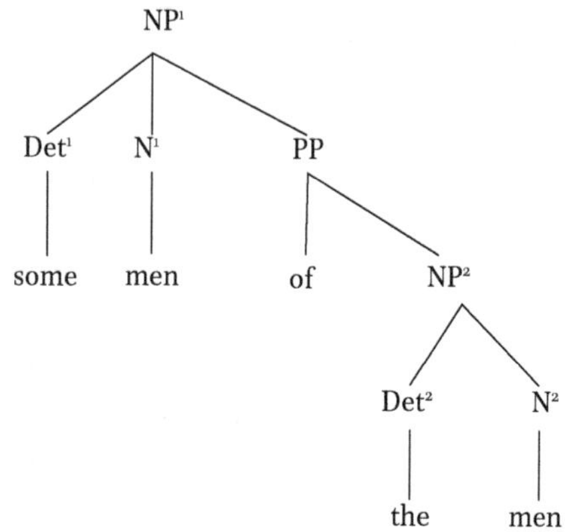

In structures of this kind, an obligatory transformation deleted N^1, provided it was identical with N^2; otherwise, the derivation blocked.[42] Still another transformation derived *some men* from the same deep structure by dropping *of* in PP as well as *men* in N^1, if Det^2 was generated as indefinite. This transformation was also obligatory and required identity of N^1 and N^2. Since both transformations were obligatory, neither of the surface structures were identical with either of the deep structures.

The second analysis, proposed by Jackendoff, was basically similar to the one proposed by Dean, except for the way N was treated lexically: on Dean's approach it was a full noun, whereas on Jackendoff's proposal it was a pronominal form:

(4.5) *every one of the men*

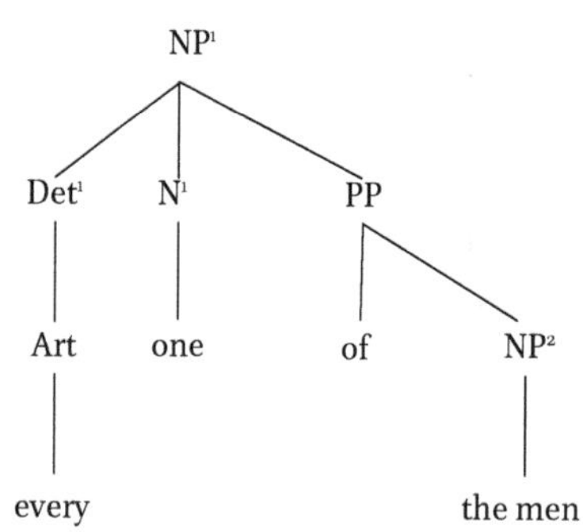

[42] J. Dean, *Determiners and Relative clauses.*

With other quantifiers, *one* in (4.5) was then deleted either obligatorily (as with *all, both*) or optionally, as in the following paradigm:

(4.6)

each
either
neither
any
which

} { man / (one) of the men }

Jackendoff also observes that the plural quantifiers like *all, some, both,* and the plural variant of *any* cannot take *ones* followed by a definite complement:

(4.7)

all
some
both
any
which

} (*ones) of the men

According to Jackendoff, structures in (4.7) with **ones* need not be explained, as *ones of the men* never occur in any context. To account, in turn, for the difference between the occurrence of *any* with and without *one* in its complement, as in *any of the men,* as opposed to *any one of the men,* he proposes a transformation, *ones*-absorption, to operate as shown in (4.8):

(4.8) *Ones*-absorption:

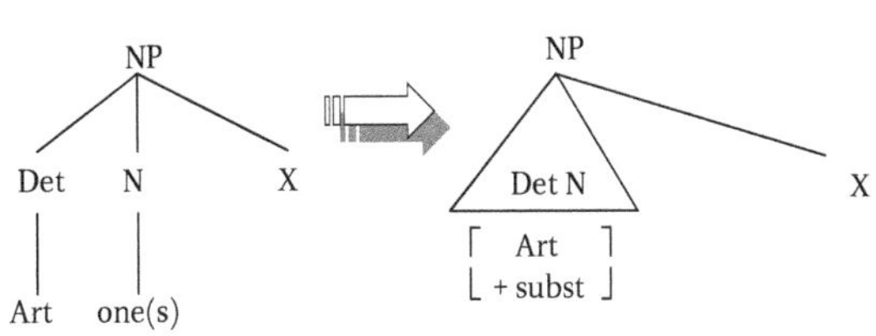

In this setting, the use of quantifiers with *one* would represent the article form, while the quantifiers without *one* would represent the pronominal form. Jackendoff goes on to point out an analogy with the difference between the use of the substantive *mine* in *This book is* ***mine*** and the adjectival *my* and *This is* ***my*** *book*. Here, according to Jackendoff, we also have the case of *ones*-absorption. The latter case is demonstrated, by Jackendoff, in the following examples. (4.9), with an adjective between the article and the noun, is acceptable:

(4.9) { no / my / John's / a / the } red one(s)

(4.10), with no such adjective, is out:

(4.10) { *no / *my / *John's / *a / *the } one(s)

Instead, *ones*-absorption produces the surface form (4.11):[43]

(4.11) none
mine
John's
one
those/it/they

What Jackendoff also demonstrates is that in (4.9)–(4.11), as opposed to the quantified cases like *any (one) of the men*, the application of *ones*-absorption turns out to be absolutely obligatory. Nevertheless, the question why, when quantifiers come into play, this obligatory character should become void seems to escape Jackendoff's attention. Instead, he gets involved—albeit to no avail—in deliberations aimed at choosing between the transformational apparatus of *ones*-absorption and a simple

---

[43] R. Jackendoff, *Quantifiers in English*, p. 440.

lexical rule systematically relating articles and pronouns, much as other lexical rules relate verbs and their nominalizations.[44] All to the opposite end, just to present quantificational cases and non-quantificational ones as just specific cases of one *general* pattern—thus pursuing essentially the goal that had already been proven unattainable, long before R. Jackendoff, by O. Jespersen in his theory of quantifiers.

4.5. *Jackendoff's Query Revisited: The Purloined Pronoun*

The main difficulty that Jackendoff encounters as well as a central issue in his treatment of quantifiers has to do with the non-existence of an appropriate pronominal form for indefinite mass nouns, in the case of which structures like (4.9) with a count pronominal form, *one*(*s*), could not be engaged, and, as a result, with a critical lack of clarity as to the constituency of a respective deep structure. To fix upon the problem, the base component of grammar should then be assumed to generate a totally abstract lexical item with no phonological shape at all, which, as Jackendoff himself notes, should be very costly in terms of an evaluation measure for theories.[45] On the face of it, however, what proves to be really at fault is the entire framework of transformational grammar as such, rather than the allegedly nebulous and inexplicit nature of the critical data that this theory strives to explain. Among the variety of the possible reasons of the failure of transformational reasoning there is at least one that has not been given due consideration yet—this reason may simply lie in the fact that the logic that actually governs the behaviour of linguistic items may happen to be entirely different from the logic that the Cartesian theorists of language would be more willing to assume.

How radically the genuine logic of linguistic facts involving quantifiers actually differs from the logic of their attempted linguistic explanation can be seen clearly if one replaces Jackendoff's transformation, "*ones*-absorption," by the same transformation only proceeding or operating in the opposite direction, say, "*ones*-extraction." Naturally, the very possibility of doing such a volte-face could not dawn upon a transformational theorist—at least not upon those who believe that the notion of deep structure as opposed to surface structure is an indispensable explicatory item in the nomenclature of contemporary linguistics. Consider, however,

[44] *Ibid.*, p. 442.
[45] *Ibid.*, pp. 441–42.

a probationary transformational grammar such that it allows the relevant strings to be generated directly at the level of surface structure, so that our transformation can proceed in the opposite direction:

(5.1) *Ones*-extraction:

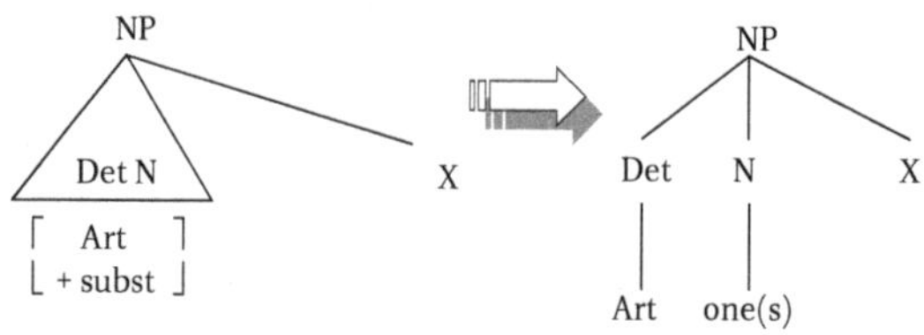

Now, the reason why there is no pronominal form for indefinite mass nouns in English would be quite obvious if we consider the occurrences of *some (one) of the men* with and without the allegedly "optional" *one*. Thus, if we consider (5.2) as an input to our provisionary transformation, and (5.3), as its output, we will readily see that they are not synonymous:

(5.2) I told three of the stories to some of the men
(5.3) I told three of the stories to some one of the men

In (5.2), *some of the men* points to a group of men while, in (5.3), *some one of the men* points to a single person, so the meaning of *some of the men* is evidently plural, whereas the meaning of *some one of the men*, singular. Accordingly, my statement that I told three of the stories to some of the men may imply that I told all of the three stories to some one of them, but not *vice versa*, which means that the corresponding statement that I told three of the stories to some man does not go to imply that they were told to anyone else at all. So, as we can see in Fig. XXIV below, the basic relational characteristics of Thirdness in (5.2) and (5.3) differ. In (5.2), where no *ones*-extraction occurs, Thirdness is relationally homogeneous, whereas in (5.3), where *ones*-extraction does occur, Thirdness is **heterogeneous**:

Fig. XXIV.

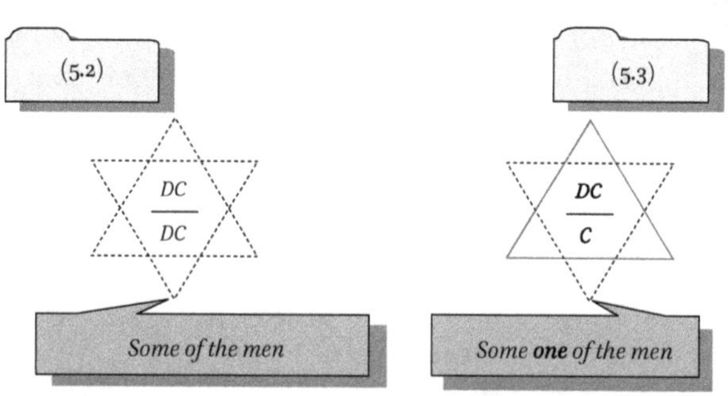

Thus, *prima facie* our transformation in (5.1)—like all other transformations when quantifiers are involved—will not be "meaning-preserving" (see Section 3.3), as the meanings of (5.2) and (5.3) are entirely different. It would then be interesting to see what the necessary semantic condition would be for the transformation of *ones*-extraction in (5.1) to be "triggered" so as to produce (5.3) out of (5.2). From inspection of Fig. XXIV we can immediately see that the relational *NON-homogeneity* of Thirdness (in bold type in Fig. XXIV) is exactly what is required to "run" the process of the "extraction" of *one* from *some*, which would then result in the structure appearing in (5.3). Extraction *would not be allowed*, however, in the case of the *homogeneity* that accompanies the plural interpretation of (5.2). Hence, a very simple explanation of the fact, noted by Jackendoff, that structures with **ones of the men*, i.e. with *one*(*s*) in the plural, never occur in any context. As is easy to see, *ones* and *the men* in **ones of the men*, should such a structure occur, would both express plurality. In other words, the referents of both *Replica* and *Token* in *the men* and in *ones* are DISCONTINUOUS, so the occurrence of *ones* is disallowed due to the homogeneity of *Thirdness* exemplified by the sameness of the quantitative characteristics of both *Replica* and *Token* in (5.2):

Fig. XXV.

QUANT + *<u>ones + of + NP</u>

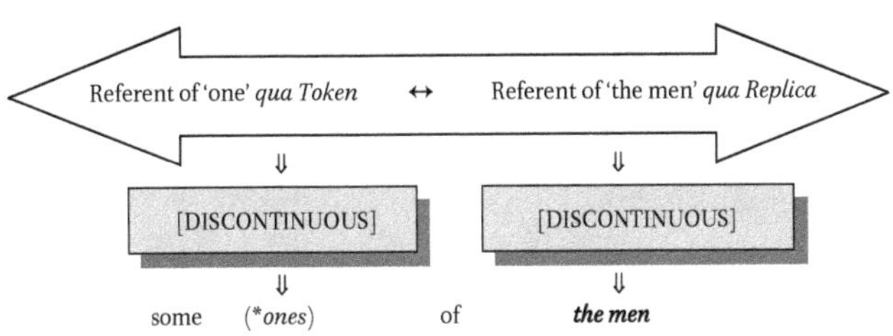

Hence, the only explanation for the occurrence of *one* is that in cases like (5.3) it turns out to be caused exclusively by the characteristics of *Thirdness*. The need for the occurrence of *one* in this case is evident, since it serves as an indispensable tool in order to signify the *NON-homogeneity* of *Thirdness*, as the quantifier alone cannot do the task when it itself is ambiguous between $\frac{DC}{DC}$ and $\frac{DC}{C}$:

Fig. XXVI.

QUANT + <u>one + of + NP</u>

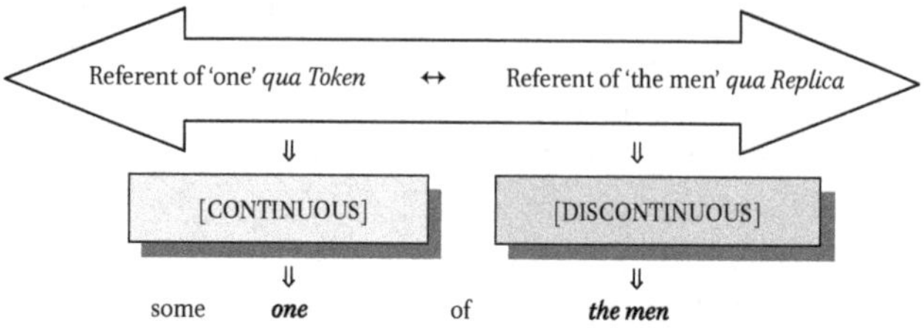

Apparently, the same structure of synthetic *a priori* connected with the relational characteristics of *Thirdness* is also standing behind the very non-existence of an appropriate pronominal form for indefinite mass nouns—the linguistic fact that appears to be most troublesome and so hard to explain in the Rationalist thinking of a transformational linguist. In order to see that this is really so, consider again Fig. XV from Section 4.1 above, which we reproduce now as Fig. XXVII:

Fig. XXVII.

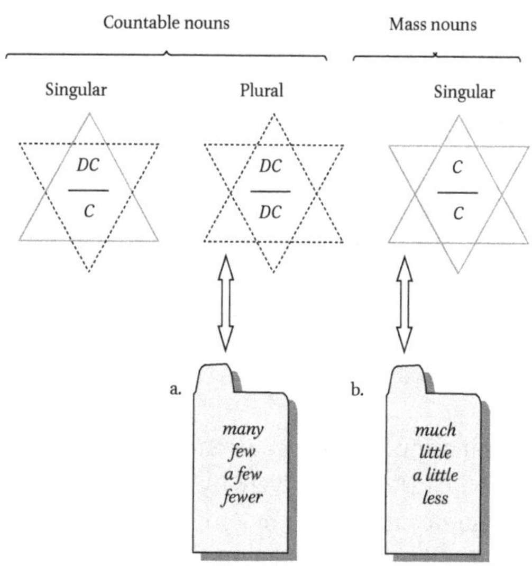

As we see from Fig. XXVII, count quantifiers like a) *many, few, a few, fewer* and their mass counterparts like b) *much, little, a little, less* both stand for the same quantitative characteristic of *Thirdness*. This common characteristic for both these groups is *homogeneity*, i.e., either the relational characteristic $\frac{DC}{DC}$ in the case of count quantifiers in (a), or $\frac{C}{C}$, for their mass counterparts in (b). So, prepositional *of*-phrases with these quantifiers appear not to require *one*(*s*) in their structure for exactly the same reason, for which plural count nouns in (a) as well as mass nouns in (b) require no article. In the latter case, the requirement that transformational structure should envisage a totally abstract lexical item with no phonological shape at all is tantamount to the requirement that deep structure should also contain an indefinite article for mass nouns. Thus, it is this context of the homogeneity of *Thirdness* that turns out to be exactly the condition that bars or rather precludes *ones*-extraction in all these cases. This is most clearly seen in the case in which mass quantifiers allow, by exception, a countable noun in the quantifier complement. The case is irregular as this concerns only the occurrence of a quantifier followed by a prepositional *of*-phrase, i.e. followed by a complex (analytic) quantifier complement, but never with a simple (synthetic) one, e.g.:

(5.4) It was fun but I'm not *much of an athlete*
much *of an athlete* ⟷ *much *athlete*
(5.5) I am indeed fortunate to see so *much of the globe*
much *of the globe* ⟷ *much *globe*
(5.6) Poor Elisabeth, she hasn't had *much of a chance*
much *of a chance* ⟷ *much *chance*

In the prepositional *of*-phrases above, the quantitative property of the referent of *Replica* is CONTINUITY. This is most vividly seen in the discourse in (5.7) in which the countable noun is replaced by a pronoun:

(5.7) "You said there was *a chance*?"
"<u>*A chance*</u>, yes; not *much of* <u>*one*</u>."

Should *ones*-extraction apply to *much of one*, the result would have been a structure like the one shown in Fig. XXVIII below:

Fig. XXVIII.

QUANT + <u>*??? + of + NP</u>

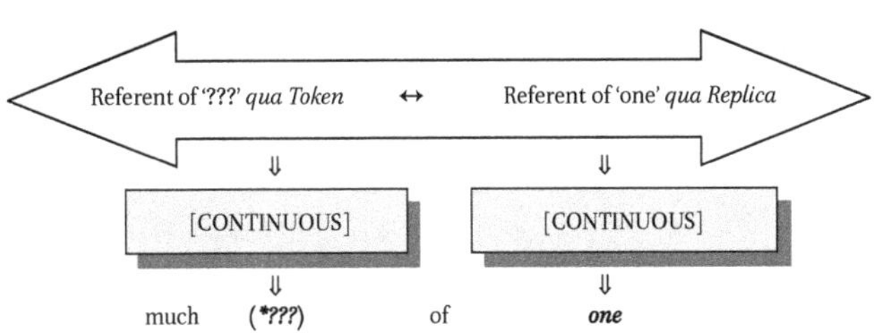

Since *much* is a mass quantifier but not a count one, the quantifier complement would now have to contain an appropriate pronominal form for indefinite mass nouns which is non-existent in English. Note that the same homogeneity resulting in the lack of any pronominal component to the left of the preposition *of* holds also in those cases when the mass quantifier is followed not by a pronoun but by a respective full noun with count interpretation:

Fig. XXIX.

QUANT + *<u>??? + of + NP</u>

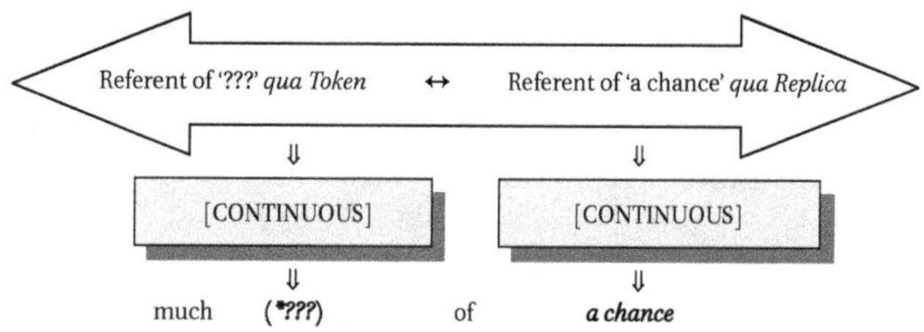

Note also that, insofar as mass quantifiers are concerned, the same holds not only for the seemingly "exceptional" cases of countable complementation, but for the regular cases of mass noun complementation as well:

Fig. XXX.

QUANT + *<u>??? + of + NP</u>

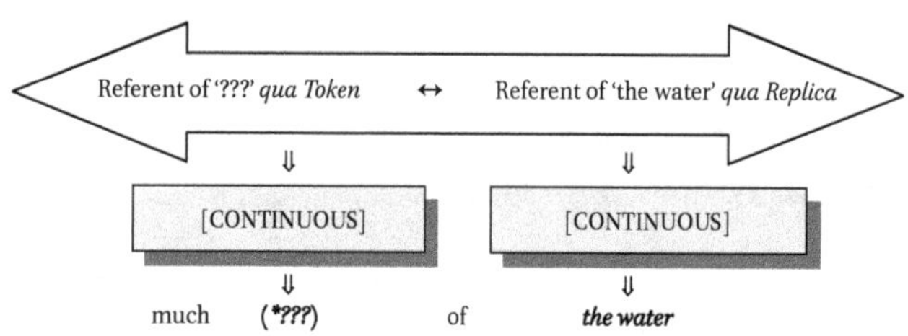

Moreover, the generality of the proposed analysis is also proved by the fact that the same pattern of homogeneity extends not only to mass quantifiers with a mass noun in place of a countable noun in the complement, but to countable nouns preceded by a count quantifier as well:

Fig. XXXI.

QUANT + *<u>??? + of + NP</u>

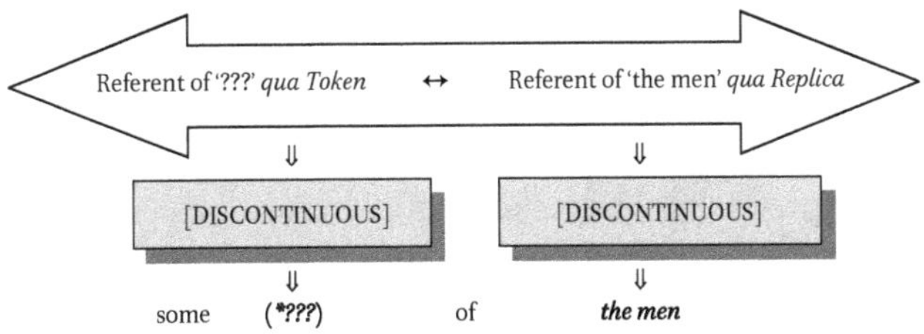

It is easy to see that the absence of the allegedly "optional" structural component in *some of the men* (due to our transformation of *ones*-extraction being blocked) takes place exactly in the case when the *plural* version of *some*, but not its *singular* version, occurs. Hence, it is exactly for this reason, namely, that the homogeneity of *Thirdness* blocks the execution of *ones*-extraction in the case of plural countable nouns and, generally, in the case of quantifiers with a $\frac{DC}{DC}$ interpretation, that structures with **ones of the men*, as Jackendoff pointed out, never occur in any context. Moreover, designating the leftmost position in the prepositional complement of the quantifier as "*X*" (that would stand either for *one*(*s*) or for the non-existent pronoun for indefinite mass nouns since they both occur in the same structural position) we would be led to conclude that our probationary transformation of "*X*-extraction" should be triggered only in those cases when it would be necessary to designate the <u>non-homogeneity</u> of *Thirdness*. Since all mass quantifiers (see Fig. XXVII) have a homogeneous $\frac{C}{C}$ interpretation (hence, belong to *Thirdness* of the $\frac{C}{C}$-type), no linguistic structure with such quantifiers will ever generate the "phonologically shapeless" *X*-element, which as said by transformational linguists they *would* generate if we should act in accordance with the logic of dogmatic speculative rationalism. However, all of the above goes to show that both of these—not only the non-occurrence of *one*(*s*) in the prepositional complement of *some* but also the very non-existence of an appropriate pronoun for indefinite mass nouns—are equally conditioned by the same synthetic *a priori* structure involving the interplay of *a priori* conceptions of quantity, *viz. continuity* and *discontinuity*.

4.6. *Jackendoff's Query Revisited: The Hidden Identity*

Jackendoff's introductory classification of quantitative expressions was but a manifestation of common wisdom offered by traditional grammar as well as to no less extent a demonstration of its weakness with regard to quantifiers. For the sake of preserving traditional classification, quantitative expressions that Jackendoff referred to as Group I words (*a group, a herd, a score, a number*) were dubbed as nouns, Group II words (*some, each, few, all, both*), as either pronouns or adjectives, and so words of Group III (*a few, many, one, three*), too. But, on the one hand, Group I words are patently far from being genuine nouns. Besides their common ability with nouns to pluralize, be counted and modified by adjectives, they also reveal certain recondite properties indicating that words of this group are fairly foreign to nouns proper: e.g., one can break *a coffee cup*, however one cannot drink it; but this happens to be precisely the opposite way with *a cup of coffee*—one can drink a cup of coffee but one cannot break it. In other words, the referent of 'cup' in 'a cup of coffee' is *coffee* but not the cup itself. Group II and Group III words, in their turn, only look as if they were adjectives, but their paradigm differs from that of adjectives proper since adjectives cannot be used as pronouns. Besides, even their adjectival use turns out to be "not truly adjectival" since Group III words do not in fact permit articles in front of them. No wonder, then, that after initially itemizing "quantifiers" as both pronouns and adjectives, Jackendoff goes on to enquire into whether they are, in the last resort, nouns or articles, or both. Desperately striving to come to terms in determining the ultimate categorial status of quantifiers in English, he however quickly reaches at the conclusion that "the available theory of grammar provides no such compromise solution."[46]

It would be appropriate here to point out that even Jackendoffs wording when he referred to "the so-called quantifiers" was not incidental. Just as Jespersen before him, Jackendoff endeavoured to discover the logic of quantifiers, the latter being understood as the linguistic counterparts of what was then known as quantifiers in mathematical logic. Jackendoff's spirit of discovery was exactly like Jespersen's, who stated that "Language is not mathematics... Language has a logic of its own" and thus clearly indicated that he regarded the exigencies of 'penetration behind the linguistic form to the logical kernel of language' as definitely distinct from

[46] *Ibid.*, p. 442.

the ordinary business of mathematical (symbolic) logic and viewed it as an eminently linguistic task.[47] For Jackendoff, the term 'quantifier' was unacceptable for the same reason, as an alien theoretical construct taken off the shelf of a mathematical logician. Instead, he viewed the exigencies of penetration behind the linguistic form as indistinguishable from the exigencies of penetration to the underlying syntactic structure of a language. Besides, for a transformational grammarian such as Jackendoff, the very use of the term 'quantifier' meant the recognition of a specific assumption of "deep structure as logical form" according to which the treatment of quantifiers in deep structure would not be wholly unlike their treatment in logic. This, of course, stood in stark contrast with both Jespersen's as well as Jackendoff's view on the treatment of quantifiers in natural language as an eminently linguistic task and, thus, as the task that is definitely distinct from the ordinary business of symbolic logic. Thus, for Jackendoff, the discovery procedure in his quest for the underlying logic of language consisted in looking for the 'kernel structures' with their 'true' way of arrangement of grammatical categories that should have been sought at an hypothesized 'underlying' syntactic level of D-Structure (Deep Structure).

The task of the so stipulated level of D-Structure was to elucidate the allegedly 'authentic' underlying syntactic arrangement of traditionally known items, so that the poorly understood phenomena of *structural ambiguity*[48] could be explained away in terms of dissimilarity of, and thus difference of form in, the 'underlying' structural descriptions. In a way, the underlying structure was no more than just syntactically disambiguated surface, or perceptible, linguistic structure. These two levels were then linked by transformations that had to rearrange linguistic items after they had been discovered in their 'genuine' syntactic positions at the level of D-Structure by moving them to alternative positions, in order for the resulting structures to square with the observable phenomena. The latter were thus conceived as only distorted manifestations of some hidden but purportedly true syntactic facts exemplifying the logic of language, the observable 'distorted' character of their surface manifestations coming as a result of the application of transformations. However, this new linguistic wisdom explained anything but the essence of linguistic structure. The hailed 'revolutionary' status of the new generative-transformationalist

[47] O. Jespersen, *The Philosophy of Grammar*, pp. 331–32.

[48] Chomsky's term in *Syntactic Structures* was 'constructional homonymity'.

theory notwithstanding, the very basics concerning the understanding of the essence of linguistic structure persisted in the same unaltered form, on which the introduction of the notion of syntactic transformations produced absolutely no effect. The very basics of this understanding still kept resorting to the idea of syntax as a mere concatenation of notional or concept-words arranged in this or that order. If the essence of linguistic structure were to be sought in the essence of such concatenation as a union in a linked series, then, what the machinery of syntactic transformations could only do by way of exposing this essence was at best pointing out at *a certain class* of distinguished unions of linked series. However, the mere setting of one union in a linked series apart from the others just to show it as the element of a hypothesized covert linguistic structure had not led by itself to any clearer understanding of its essence whatsoever. Nevertheless, what in Jackendoff's quest for the hidden identities, sought after in the underlying structures involving quantifiers, has been proven to be really feasible was whatever result, however *not* even to that limited extent. All for one reason: in his research on the syntax of quantifiers in English he had been guided by the philosophical principle of apprehending essence as form, pervasive in linguistic and logical thought ever since Aristotle. The principle that required the need for attaining isomorphism between syntax and semantics as a categorical imperative. As a corollary, syntax was foreordained to be perceived as something sublime, primary, primordial, therefore wholly self-sustained and completely autonomous, while semantics, as its mere reflection and, at best, as something manifestly derivative. However, the use of Peirce's categories in the undertaken in-depth analysis of phrasal quantifier structures—that had been left unexplained in Jackendoff's attempt; moreover, wholly abandoned as a subject matter and, to all intents and purposes, totally ignored in the subsequent explorations at the level of LF (Logical Form) as a level of linguistic representation—has clearly shown that this principle is fundamentally flawed, and the ensuing conception of autonomous syntax, demonstratively false. Moreover, what the analysis in question has in fact revealed is that it is exactly abiding in this principle, which is overtly heading general linguistic inquiry exactly in the opposite direction, which keeps the actual truth effectively disguised.

What we can clearly discern now are the two vividly observable and *prima facie* detached yet inextricably connected patterns of failure, the first one stemming in the highly controversial issue related to the notorious failure of transformations to preserve meaning in the presence of quantifiers, and the second, connected with the failure to trace the phrasal quantifier structures down to their ultimate syntactic constituents. However, it is not merely the presence of quantifiers that makes these patterns grossly related. Notably, it is the very fact that the failure of transformations to preserve meaning just in the presence of quantifiers spells paradox that turns out to be, in the last analysis, but a mere consequence of a much less discussed failure to deal with the challenges that the ultimate analysis of the phrasal syntax of quantifiers present to the theorist.[49] The latter failure, however, did not come as a result of the lack of the appropriate 'categorial' means in the available theory of grammar, as Jackendoff seems to be suggesting, but was merely destined to come about as a straightforward and instant corollary of precisely the idea, or principle, of "autonomous syntax."

Given the thesis of "autonomous syntax," the upshot of the analysis at the syntactic level of D-Structure was the generation of the sequences of symbols such that each of these sequences would always be uniquely interpreted. Naturally, such a condition could only be fulfilled provided that the symbols and their configurations generated by the base component would stand in a strict one-to-one correspondence with certain external, generally, semantic entities in their respective configurations. However, given the explanatory power of Peirce's categories, as shown by the undertaken analysis of the phrasal syntax of quantifiers, this condition simply cannot be met. More specifically, this condition would have only been met if, for individual meaningful symbols or units of symbols that Phrase Grammar had been assumed to generate, Peirce's Secondness could have further been thought of as a category of the uppermost complexity. Yet, due to the very fact that any meaningful expression (from a single noun to a functional expression in Frege's logical notation) shares the being in the two modes and thus exists simultaneously as both Token and Type in one, and therefore its genuine description just cannot be restricted exclusively to Secondness as the upper limit of existential complexity but necessarily entails and should therefore, of necessity, imply a yet higher level of existentiality, that of Peirce's Thirdness,

[49] As noted by one of the authors of the Lakoff-Carden "logical" approach to the linguistic analysis of quantifiers that initiated, back in the late 1960's, a total divorce from the specific issues of the syntax of phrasal quantifier constructions, indicative of all further development of linguistic thought up to the present moment: "The phenomena I have considered do not force me to distinguish between partitive and non-partitive use of quantifiers ("many men" vs. "many of the men"); I regard the correct solution for this problem as an open question." (G. Carden, *English Quantifiers: Logical Structure and Linguistic Variation.*, p. 1).

any idea of attaining any strict one-to-one correspondence between syntactic and semantic entities, even in theory, is just untenable. As a corollary, no theory of analytic *a priori* 'underlying' level of syntactic structure (be it any theory like that of D-Structure or LF-theory), nor any theory resorting to whatever sort of 'underlying' *quasi*-logical symbolism can ever be tenable, either.

CHAPTER FIVE

GÖDEL'S INCOMPLETENESS THEOREM AND THE DOWNFALL OF RATIONALISM: VINDICATION OF KANT'S SYNTHETIC *A PRIORI*

But now, if the misunderstood Kant has already led to so much that is interesting in philosophy, and also indirectly in science, how much more can we expect it from Kant understood correctly?

Kurt Gödel, Collected Works

Heisenberg, Gödel and Chomsky walk into a bar. Heisenberg says "This is very odd and improbably, and I wonder if we might be in a joke, but I can't be certain." Gödel says "Well, if we were outside the joke we would know, but since we're inside the joke, there's no way of determining whether or not we're in a joke." And Chomsky says "Of course this is a joke, but you're telling it wrong!"

John M. Ford

As Weber said, you can't look at the components of history in isolation. Just as you can't see the Reformation and the rise of capitalism separately, so you can't see the degeneration of philosophy, from the dialectic of ideas to the endless parenthesis over words, and the rise of mass media separately.

It's all about words, because words are the stock-in-trade of mass media, which rules the world. Thus, characters like Russell, Wittgenstein, Derrida, Foucault, Barthes, Chomsky and their fellow brigade of linguistic/analytic, deconstructionist, post-modernist, post-post-modernist nihilists, have achieved totally undeserved cachet. The mischief started with the notion that instead of language serving meaning, it's the other way around.

Anonymous commentator @www.jewcy.com

5.1. *Chomsky's Understanding Understanding and Gödel's First Incompleteness Theorem*

On arithmetical ability, Chomsky says:

> Take the human number faculty. Children have the capacity to acquire the number system. They can learn to count and somehow know that it is

> possible to continue to add one indefinitely. They can also readily acquire the technique of arithmetical calculation. If a child did not already know that it is possible to add one indefinitely, it could never learn this fact. Rather, taught the numerals 1, 2, 3, etc., up to some number *n*, it would assume that that is the end of the story.[1]

This passage serves well as a concise pronouncement of Chomsky's calculus views on language. Here, "number" and "the number system" are freely intersubstitutable with "language" to produce the equally valid "Take the human language faculty. Children have the capacity to acquire language." And the rest goes to explain the very basics of Chomsky's linguistic turn, with "the end of the story" being the key words to explain the conundrum.

"The end of the story" here clearly refers to Chomsky's famous predecessor Leonard Bloomfield, whose Empiricist theory of descriptive linguistics explained language acquisition in empirical terms of stimulus-response theory. According to this theory, a child only understands sentence structures that it recognises from previous experience in much the same way as it knows lexical items the meanings of which have previously been given to it by ostensive definition. Thus, a person would never have understood the meaning of some number *n+1* nor the meaning of any acceptable sentence in English if it had not belonged to the person's previously acquired stimulus-response patterns.

Speaking about the number series *1,2,3,.. n* and also about a child knowing that it is possible to add one indefinitely, Chomsky explicitely refers to the axiomatic system of Frege's *Grundgesetze* that he puts in the very basis of his Rationalist theory of language, which he propounded as a means to resolve the evident paradox of a manifestly Empiricist behavioral theory of language development. In much the same way as Frege before him modified mathematical function theory to suit the needs of his new system of logic, Chomsky adapts Frege's axiomatic system of logic to meet the specific ends of his structural revolution in linguistics. Instead of Frege's concern at how to demonstrate that a novel mathematical statement is true, Chomsky's newfangled concern is how to make known the way one understands or produces a sentence that he never met with in prior experience or was never taught how to construct. Since Frege and Chomsky firmly stand on the grounds of Rationalist philosophy, no wonder that

[1] N. Chomsky, *Language and the Problems of Knowledge*, p. 167.

the intellectual tool that they both use and the purpose to which it serves happen to be eventually the same. For Frege, the purpose is to show how any of an unlimited number of new mathematical truths that he calls theorems unfold from a short list of basic elementary truths that he calls axioms. The tool to serve this purpose is a system of inference rules that basically are rules of production capable of generating new statements out of axioms and statements resulting from previous derivations.

For Chomsky, on the other hand, the purpose is to spell out how new sentences in a language unfold from a short list of elementary phrase structures. The tool to serve this purpose is a system of phrase structure rules called syntactic transformations that are essentially the kind of Frege's production rules that apply to the basic syntactic patterns and the sentences already derived from them to reach at a larger number of follow-up sentences. Thus, on a strict analogy with mathematical number theory, the theory of child's understanding of statements about new mathematical objects such as numbers smoothly transforms into the theory of understanding in general.[2] Therefore, since a child cannot know in advance all the sentences in a language that it confronts, what the child needs to know instead are respective production rules (by analogy to that of adding *one* indefinitely) that it would then recursively apply to all sentence structures that are new to the child so as to understand them, in much the same way as a mathematician applies his inference rules in order to recognize the truth of mathematical theorems. Accordingly, all Chomsky had to do was to posit the logical notions of "truth/analytic truth," and the linguistic notions of "acceptability/grammaticality" mutually interchangeable. Apart from these minor differences of detail, both Frege's axiomatic system and Chomsky's system of transformational-generative grammar appear in fact as no more than only twin extensions of the very same notion of formal system.[3]

From Frege's viewpoint, if a statement of number theory is true then it is provable, i.e., derivable from the axioms strictly and mechanically via the application of the rules of inference. Accordingly, if a statement is provable, it is true. In this way, "truth" and "provability" look like the

[2] ...in much the same way as Frege before him expanded the notion of function so as to include, as a value of the argument x in the functional expression, $f(x)$, any other objects besides numbers, see p. 11 above.

[3] Cf. Willem J. M. Levelt, *An Introduction to the Theory of Formal Languages and Automata*; Jerzy Bańczerowski, *The axiomatic method in 20th-century European linguistics.*

two sides of the same coin and and very much remind us the two expressions, *the morning star* and *the evening star*, as just two different ways of denoting one and the same object, *Venus*. Analogously in Chomsky's view, if a sentence in a language is acceptable and could be understood by a native speaker then it is derivable in terms of a transformational grammar which is part of the linguistic competence of an ideal speaker-hearer. And, accordingly, it is the derivability that is central to the understanding of a sentence since, as in Chomsky's cases of "constructional homonymity," different understanding of identical structural representations are only explained away by their different "derivational histories." Thus, according to Chomsky, understanding a difference between the two meanings of, say, *the shooting of the hunters* amounts to knowing its transformational history, i.e., to knowing whether it is derived from *the hunters shot someone* or from *someone shot the hunters*. Thus, the Rationalist mathematical wisdom that the truth of a number-theoretic statement and its provability coincide, and are thus inseparable like the two sides of a coin or like the front and back of a piece of paper, gets elaborated into its mirror-like equivalent in the wisdom of a Rationalist linguist, that understanding the sentence equals understanding the sentence's transformational history. This is so much so in view of the fact that what you would only be able to understand is a sentence (in English as in any other language), that should necessarily be grammatically correct. Thus, a sentence is acknowledged to be grammatically correct (thus, acceptable) if and only if it has been derived from base-generated structures by transformational rules, in exactly the same way as a number-theoretic statement is known to be true if and only if it has been shown to have been derived from axioms by inference rules.

This obvious parallelism is apt to readily invite interesting questions, though. The rise of Chomsky's transformational-generative grammar dates back to 1955, which was long after a dramatic event in the history of the theory of formal systems, widely known as the proof, by Gödel back in 1931, of his famous Incompleteness Theorems. These theorems are notorious for striking a serious blow to the credibility of the very notion of formal systems. As a person that was well read in both finite automata theory and recursive function theory, Chomsky was well aware of it. Yet, he pressed the gas pedal while the yellow light was already on. Why?

As James Kelman notes in his essay *A reading from Noam Chomsky and the Scottish Tradition in the Philosophy of Common Sense*:

> Chomsky was well aware of Gödel's Theorem. Even back when he accepted the orthodox view of linguistics—that semantics should be excluded

> from the study—he had his own distinctive approach: a linguistic theory should not be identified with a manual of useful procedures, nor should it be expected to provide mechanical procedures for the discovery of grammars... We cannot hope to say whether a particular description of the data is correct, in any absolute sense, but only that it is more correct than some alternative description of the same data... The most that can be expected is that linguistic theory should provide criteria (an evaluative procedure) for choosing between alternative grammars.
>
> In comparison to the Bloomfield/Harris objective, as John Lyons points out, Chomsky's objective here seems quite unassuming, but ultimately it is more ambitious. Einstein's physical system is greater than Newton's because it is more powerful, it copes more adequately with the raw data of the universe. His system can do what Newton's can do, but it can do a great deal more. Yet nowadays we know enough about systems in general to appreciate also that the Einstein version is not the last word, not in any absolute sense. Eventually another system will come to supercede it. Once this point is realized Chomsky's ambitions become clearer, he is seeking a form of ultimate criteria, universal principles by which different grammars may be evaluated.

Here, again, Chomsky's ambition is nothing more and nothing else than a desire to accommodate the idea of a model-theoretic approach and that of a metatheory as a means of evaluating various systems in logic to the evaluation of diverse applications of the idea of formal system to linguistic structure. Just as Chomsky's refusal to state whether a particular description of the data is correct, in any absolute sense, follows simply from his awareness of the metatheoretical implications of Gödel's Theorem, and also of the fact that no metatheoretical approach can provide any cure against the negative results of Gödel's Theorem, since the latter is itself a statement of metatheory. However, Chomsky's view that "a linguistic theory should not be identified with a manual of useful procedures, nor should it be expected to provide mechanical procedures for the discovery of grammars" deserves a bit closer reflection. For it may be grounded on Gödel's own distinction between "subjective" vs. "objective" mathematics or on Lucas' view "that there is some elusive and ineffable quality to human intelligence" which no machine can duplicate.

Lucas' view is, however, trivially incorrect, as he draws his contention upon something that Gödel's Theorem does not state. Thus, Lucas takes it for granted that a sentence σ such that neither σ nor its negation $\neg\sigma$ is provable in P can immediately and self-evidently be known to be true:

> However complicated a machine we construct, it will, if it is a machine, correspond to a formal system, which in turn will be liable to the Gödel procedure for finding a formula unprovable-in-that-system. This formula the

> machine will be unable to produce as true, although a mind can see that it is true. And so the machine will not be an adequate model of the mind.[4]

In Gödel's Theorem, the truth of σ is only presumed by the initial assumption that P is consistent, and it does not mean anything else than just that. For nowhere does Gödel state that, by a human mind, the truth or falsehood of σ may be somehow known in advance. Quite on the contrary, he treats σ only as a question that requires an answer, and the only possible way to obtain such an answer is to prove σ as a theorem in P:

> But, on the other hand, one clung to the belief... that every precisely formulated yes-or-no question in mathematics must have a clear-cut answer. I.e., one thus aims to prove... that of two sentences A and ~A, exactly one can always be derived. That not both can be derived constitutes consistency, and that one can always actually be derived means that the mathematical question expressed by A can be unambiguously answered.[5]

Thus, insofar as purely mathematical cognition is concerned, what a human mind can only know in advance is that a sentence, σ, is actually a yes-or-no question the answer to which will be "yes" if σ can be derived, or "no" if $\neg\sigma$ can be derived from the axioms. Thus all that the human mind can only know prior to any proof procedure is only that σ should necessarily be either true or false but not both. On Frege's analytic model of mathematical knowledge, what can be instantly known to be true are axioms. Whatever knowledge that can count as new is wholly contained in the axioms but, in order to be directly accessible, it must first be "unfolded" from the axioms in much the same way as a plant grows (and thus "unfolds") from its seed. Just as we cannot see a tree when we look at its seed, knowledge of anything other than axioms cannot be obvious or else directly perceived by the mind; its truth cannot be instantly recognised at the very first glance of the "mind's" eye, it can only be recognized indirectly via its demonstration by proof. Since all knowledge in (discrete) mathematics is analytic, this means that there is (and there can be) no new mathematical knowledge besides that which was previously covertly contained in the axioms, just as there can be no plants besides those that, previously, were seeds. Thus, no new (mathematical) knowledge can be true in any obvious way, it can only be demonstrably true—that is, it can only be proven to be true by demonstrating whether and how it can be

[4] J. R. Lucas, *Minds, Machines and Gödel*, p. 48.

[5] Kurt Gödel, *The modern development of the foundations of mathematics in the light of philosophy.*

derived (produced) from the axioms. Thus, Lucas' contention that there can be a "formula the machine will be unable to produce as true, although a mind can see that it is true" is, insofar as mathematical knowledge is concerned, demonstrably false.

But, over and above that what Gödel's First Incompleteness Theorem intrinsically presupposes, most essential is what it overtly articulates. And, to use Gödel's own words, what this theorem states as a proven fact is that

> Even if we restrict ourselves to the theory of natural numbers, it is impossible to find a system of axioms and formal rules from which, for every number-theoretic proposition A, either A or ~A would always be derivable.[6]

Just as Kant's *Critique of Pure Reason* states as a proven fact that *human* knowledge has limits, Gödel's Theorem proves essentially the same, namely, that *mathematical* knowledge has limits. What all of this effectively means is that a mathematician can pose more questions than he can answer, and the questions that will remain unanswerable are exactly those σ's that, in P, cannot be proven. This in turn leads Gödel to distinguishing between what he terms as *objective* and *subjective* mathematics. Since the mind of a mathematician is a kind of Turing machine (and it cannot be otherwise unless mathematical knowledge is non-analytic), all the knowledge that he is able to have access to (all the theorems that are provable in P) is henceforth *subjective*, as opposed to all knowledge (regardless of whether or not it is accessible via proof procedures) that is independent of the cognizing mind, hence, *objective*. Thus, in Gödel's terms, *objective mathematics* is just another name for absolute knowledge, or else knowledge from a God's eye point of view, of the truth or falsehood of all of the theorems in P (since P is consistent), including the truth or falsehood of those theorems that cannot be proven.

In Chomsky's terms, this would translate to a corresponding distinction between *subjective* and *objective* linguistics. Since what generative-transformational linguistics is wholly about is the linguistic *competence/performance* of an ideal speaker-hearer, it just cannot be about something that this speaker-hearer is not competent to perform, and in this sense it cannot be anything but *subjective*. Hence, the respective counterparts of the undecidable statements that a Turing machine cannot deal with simply

[6] *Ibid.*

do not exist. Therefore, "a linguistic theory should not be identified with a manual of useful procedures" since the ideal speaker/hearer's mind need not be identified with a Turing machine, as the former need not have to cope with anything that the latter was intended to be able to cope with, but turned out to be incapable of doing it. For, in Chomsky's linguistics, "[n]o 'Platonism' is introduced, and no 'E-linguistic'[7] notions: only biological entities and their properties."[8] Thus, to cite a pro-Chomskyan linguist, "the status and interpretation of Gödel's results have no particular bearing on linguistics."[9]

Chomsky's viewpoint, however, happens to be fundamentally flawed, as there is more going on in Gödel-Chomsky connection than meets the eye. In what follows, the point to be made is that Gödel's Incompleteness Theorem, just in view of Chomsky's escape route into the safe haven of biolinguistics, demonstrated much more than Gödel might have thought it actually did. For even if it is not immediately evident, it is nonetheless obvious that Gödel's First Incompleteness Theorem had in fact vindicated Kant's stance that all significant mathematics depends essentially on a non-logical basis, our *a priori* forms of intuition, by showing, since it is not possible to arrive at an analytic result from a synthetic base, even when each step along the way is analytic, that mathematics is ultimately synthetic at its roots, and thus synthetic at its leaves.

5.2. *Gödel, Chomsky, and the Synthetic Base of Mathematics. Part I*

As Emmon Bach testifies,

> What was most original in Chomsky's early work was the very idea of a generative or formal grammar, considered as a theory that would specify all and only the (infinitely many) expressions in a language, and assign structural descriptions to them. In a kind of Kantian turn, linguistic theory became then the study of the general structure of the grammars that would be just adequate to capture natural languages as formal systems. The idea that a language could be described as a formal system might well be called "Chomsky's thesis," even though, as is often the case, the idea was "in the air" at the time (Bar-Hillel, Harwood, Greenberg, Hockett, among others).[10]

---

[7] "E" is to suggest 'external' and 'extensional'.
[8] N. Chomsky, *Derivation by Phase*, p. 42.
[9] John Collins, *A Question of Irresponsibility: Postal, Chomsky, and Gödel*, pp. 102–3.
[10] Emmon Bach, *Linguistic Universals and Particulars.*

From the foregoing, we can see that "Chomsky's thesis" was the idea to specify all and only the (indefinitely many) expressions in a language, captured as a formal system, where "all and only" plays a crucial role, since the specific use of this expression, 'only', is clearly intended as a safety lock to guard against the negative consequences of Gödel's Theorem. To see how it works, consider the expression 'an expression in a language', or shorter, 'an expression in L' as an exact formal equivalent to an expression in a number theory, 'a theorem in P'. These will then be derived by production rules that are, respectively, transformational rules (transformations) in the formal theory of language, and inference rules in number theory, from the respective initial strings that are base-generated structures in generative-transformational grammar, and axioms in an axiomatic system of number theory. Note that 'generative-transformational grammar' and 'formal theory of language' here is taken to mean basically the same. Essentially, both (an axiomatic system of) number theory and generative-transformational grammar share the same common components:

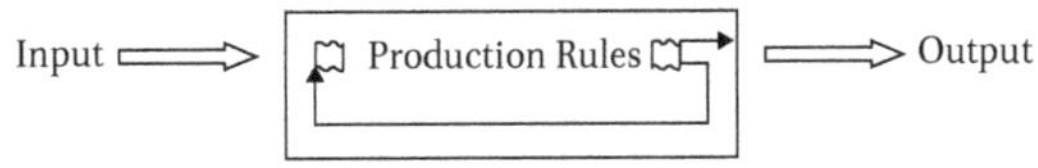

Since production rules can operate not only on input data, but also on their own output (as indicated by the arrows above), the process can go on indefinitely long, and thus produce an indefinite amount of new data with no bound, which are either grammatically correct (acceptable) expressions in L, in one case, or valid theorems in P, in another. The relevant difference here is that in the linguistic case, the formal system does indeed produce only grammatically correct expressions, and thus seems to be able to produce all and only grammatically correct expressions in L. In the number-theoretic case, it does produce only valid theorems (since the negations of invalid theorems that it can also produce are valid theorems, too). That it can produce only valid theorems, but *not all of* them, is a mathematically proven fact due to Gödel's First Incompleteness Theorem.

Now, let us consider if a formal system in its linguistic version does present itself as an exceptional case of a formal system, exceptional in that, in some way or another, it is immune to Gödel's Theorem. To show that this case holds, we will need to present a proof that the formal system in its linguistic disguise not only seems to be able to produce all and only grammatically correct expressions in L, but does indeed produce *all* of them (that it produces *only* grammatically correct expressions we need not prove since this is trivially evident). But, delivering such a proof is tantamount to nothing more and nothing less than to proving that Gödel's First Incompleteness Theorem is, in fact, invalid. Since the latter, however, is a mathematically proven fact, the former is impossible. Hence, whatever substantiation of the thesis that this system does produce not only those expressions in L that are grammatically correct, but also *all* grammatically correct expressions in L, it will by logical necessity be based on a logical fallacy.

That some such consideration would inevitably involve some kind of sophistry and opulent talk can be easily seen from the imbroglio between Paul Postal and Chomsky, in which the former accuses the latter of doing junk linguistics. In the relevant passage, cited by Postal, Chomsky pursues the issue of ontological commitments:

> These [that is, expressions, the language faculty outputs—PMP] are not entities with some ontological status; they are introduced to simplify talk about properties of [the language faculty], and they can be eliminated in favor of internalist notions. One of the properties of Peano's axioms PA is that PA generates the proof P of '2 + 2 = 4' but not the proof P' of '2 + 2 = 7' (in suitable notation). We can speak freely of the property 'generable by PA', holding of P but not P', and derivatively of lines of generable proofs (theorems) and the set of theorems without postulating any entities beyond PA and its properties.[11]

Postal, however, does not seem to be getting the point that Chomsky is actually making, and speaks, instead, about something that Chomsky is not talking about at all, not in this passage at least. Postal erroneously assumes that in this passage Chomsky makes "his truly desperate claim ... that *sentences are not real,*"[12] and concludes his argument stating that "Chomsky's claim about 'without positing any entities' is absurd."[13] The equivocation here arises, however, from Postal's misinterpretation of the statement that *sentences are not real,* which he takes to be a universal one, i.e., *all sentences are not real,* whereas the point that Chomsky is really making shall be properly rendered, rather, as a corresponding

[11] N. Chomsky, *Derivation by Phase,* p. 41.
[12] Paul Postal, *The Incoherence of Chomsky's 'Biolinguistic' Ontology,* p. 117.
[13] *Ibid.,* p. 118.

particular statement *some sentences are not real*. The key issue that Chomsky here raises is that

> One of the properties of Peano's axioms PA is that PA generates the proof P of '2 + 2 = 4' but not the proof P' of '2 + 2 = 7' (in suitable notation).

Therefore, he speaks here of the non-existent proof of '2 + 2 = 7' which means that '2 + 2 = 7' is not a theorem of PA. Here, what Chomsky speaks of as *not real* is the non-existent proof of '2 + 2 = 7' which is non-existent in the absolute sense, and of the statement that "it is true that 2 + 2 = 7" which, despite its being existent as a false statement, is non-existent as a theorem in PA, but not of anything over and above that. Here, however, an equivocation arises on Chomsky's own side, since what makes Chomsky's argument ultimately invalid is the equivocation between 'without positing any entities' *in general* and 'without positing any entities *beyond PA'*, i.e., between being non-existent in the *absolute sense* and being non-existent in the *relative sense*, when something does really exist but the existence of this something is irrelevant for the purpose at hand and thus need not be accounted for. Thus, what Chomsky means by

> We can speak freely of the property 'generable by PA', holding of P but not P', and derivatively of lines of generable proofs (theorems) and the set of theorems without postulating any entities beyond PA and its properties.

is that we can freely speak about grammatically correct a.k.a. acceptable expressions in L as generable in L without bothering about the existence of any expressions beyond those that are generable in L, because even if they do exist, such expressions are grammatically not correct, hence they are not expressions in L, thus for the purpose at hand the very issue of their existence or non-existence is utterly irrelevant.

Although true, this is beside the point. In Chomsky's exposition, the fact that *PA generates the proof P of '2 + 2 = 4' but not the proof P' of '2 + 2 = 7'* goes to show that, since *the proof P'* is non-existent, the corresponding entity, '2 + 2 = 7', is to all intents and purposes irrelevant as it appears to belong to the class of entities that go *beyond PA and its properties*. The relevant question, however, is not whether '2 + 2 = 7' exists or not, nor is it a question whether we can or cannot do without postulating or positing it as a entity, but rather if '2 + 2 = 7' can be answered when it is posed as a question. In this latter sense, to wit, neither AP nor its proofs taken in themselves have any ontologically self-subsisting, independent, or else Platonic status, either. They are only tools *in the hands of an investigator* (mathematician) and only exist, to put it in Heideggerian terms, as a piece of equipment that is 'ready-to-hand', as a man-made artifact that serves to attain certain goals and thus exists only insofar as it serves its intended purposes.

Viewed now from Gödel's own perspective, all the business that PA and its properties have to attend to, in the case of '2 + 2 = 7' (in suitable notation), is answering the question if P' is decidable or, posed as a yes-or-no question whether P' is true or not, answerable. Then, what the non-existence of the proof P' of '2 + 2 = 7' tells us is that P' is semi-decidable, and that is only half of the story as what it means is only that the question if P' is true or not cannot be answered in the affirmative. But having gained this result we are not yet in a position to state whether the question is answerable as all that we know by now is only that it cannot be answered in the affirmative, yet we do not know yet if the answer can be in the negative. The answer will be provided if the respective proof, namely *the proof ¬P'*, is found. If it does not exist, the proof will never be found, and the question will remain forever unanswerable. In each particular case, however, we cannot know in advance whether this or that question posed by a mathematician *can* be answered. Nor Gödel's First Incompleteness Theorem can provide an answer to this question. What it only says is that in this or that particular case this question may be unanswerable, and it in fact may be the case that the theorem in question just cannot be solved in principle. Thus, Chomsky's argument that "we can speak freely without postulating any entities beyond PA and its properties" simply loses sense, and it does so not only because we will, contrary to Chomsky, necessarily need to have to posit *P'* "as an entity" even despite the non-existence of the respective *proof P'*, but equally, and even more importantly, because *the set of theorems* and *the set of theorems as generable proofs* that Chomsky thinks are equal, due to the very existence of *unsolvable theorems, are not*, and cannot be, coextensive. Thus the equivocation leading to the totally unsupported conclusion that "the status and interpretation of Gödel's results have no particular bearing on linguistics" appears to rest on Chomsky's indiscriminate use of the *set of all theorems* and *the set of all provable theorems* as synonyms. What this all amounts to is that Chomsky (perhaps inadvertently) employs a classically circular argument in which the conclusion is already contained in the premises.

All these misconceptions notwithstanding, Chomsky's thesis (in the sense of Emmon Bach's definition quoted above, p. 192) appears, however, to be in line with the effort of other philosophers, concerned with the mind *vs.* formal system problem. The apparent mist of uncertainty surrounding the enquiry into the powers of human reason and its limits resulted in the urge to search for the answers in various quarters, and Chomsky was not alone in his expectation to find the answers to the problems of mind in biophysical or neurological sciences. Even Gödel, despite his views concerning the "impossibility of physicochemical explanations of... human reason," grants the possibility

> ... that brain physiology would advance so far that it would be known with empirical certainty
> 1. that the brain suffices for the explanation of all mental phenomena and is a machine in the sense of Turing;
> 2. that such and such is the precise anatomical structure and physiological functioning of the part of the brain which performs mathematical thinking.[14]

This tendency can be well understood, since it comes as a corollary of Gödel's discovery of the incompleteness of formal systems that we could not have *mathematical* evidence for the adequacy of any formal system. It would then be natural to expect "the possibility... that we have some kind of *nonmathematical* evidence for the adequacy of such."[15] In itself, however, it is an obviously self-contradictory statement, since whatever kind of nonmathematical evidence for the adequacy of any formal system can ever be found, it can only be so done on the assumption of its inadequacy, which is absurd. Rather, in order to bypass inevitable blunders and thus provide an adequate basis for dealing with the problem, one should instead seek strictly *mathematical* evidence that, appearing in a seemingly *nonmathematical disguise*, may be missed by casual view. To accomplish this, we need to abandon barren ontological issues in favour of raising concurrent epistemological issues, surfacing those related to ontology, in the metaepistemological spirit of Kant's transcendental philosophy.

5.3. *Gödel, Chomsky, and the Synthetic Base of Mathematics. Part II*

To show that Gödel's Incompleteness Theorem has in fact demonstrated that no fully satisfactory purely analytic account of mathematics can ever be forthcoming, and that one of its implications is that, over and above Lucas' contention that, in mathematics, "synthetic *a priori* truths are possible,"[16] they are, indeed, necessary, let us consider, along with

---

[14] Kurt Gödel, *Collected Works, Vol. III: Unpublished Essays and Lectures*, p. 309.
[15] Georg Kreisel, *Which number-theoretic problems can be solved in recursive progressions on Π^1_1 paths through O?*, p. 322.
[16] J. R. Lucas, *The Implications of Gödels Theorem*, p. 223.

the common formal system-theoretic background of the two different instantiations of formal system in question (those in mathematics and in linguistics), also their respective concomitant, but not necessarily coinciding, epistemological aspects.

The necessity of this shift in focus is endorsed by the simple fact that—viewed against the background of Kant's transcendental philosophy—disputes biased on purely ontological issues either in linguistics (like the one between Postal and Chomsky) or in the standard theory of formal systems (whether unsolvable problems are kind of Platonic entities to which the mind has no access) are essentially beside the point. That these discussions are irrelevant is not immediately apparent, though; since what makes the interpretation and evaluation of Kant's transcendental idealism so difficult is that it is best viewed as an *alternative to* ontology, rather than, as it usually is, as an alternative *ontology*.[17] As Allison points it out,

> Given the way in which Kant draws the contrast between the two forms of transcendentalism [i.e., transcendental realism *vs.* transcendental idealism—AM], effectively viewing them as all-inclusive and mutually exclusive alternatives, it follows that transcendental idealism must likewise be seen as a metaepistemological position, committed to *an alternative model of cognition,* and *not as a competing metaphysical theory* [emphasis mine—AM]. Otherwise they would not conflict with one another in the way in which Kant clearly assumed that they do. Moreover, since the contrast is with the theocentric paradigm, the paradigm appealed to by transcendental idealism must be anthropocentric. In short, the conditions of human cognition, whatever they may turn out to be, rather than the unattainable ideal of a God's-eye view of things, determine the norms of our cognition.[18]

From this it naturally follows that the total failure to assess fully the ultimate philosophical relevance of Gödel's Theorem as a formal model of the mind arises from the inability to differentiate between the objectivity of our representations of things from the very existence of the things themselves. As a model of (mathematical) cognition, formal system theory thus turns out to be unattainable (which is exactly what Gödel's Theorem in fact glosses in formal terms) to the very same extent to which unattainable is the ideal of a God's-eye view of things. The ideal of a God's-eye view of things is exactly the epistemological position of philosophical Rationalism. Hence, abandoning the inevitable theocentric view of philosophical Rationalism and taking on, instead, the anthropocentric "Copernican" view of

[17] Henry E. Allison, *Kant's Transcendental Idealism*, p. 123.
[18] *Ibid.*, p. 114.

Kant's philosophy gives us a far better view on formal system as a model of cognition reflecting the structure of the mind rather than the nature of a pregiven reality. Naturally, this provides a more adequate bases for dealing with the problem whether minds are machines (or, accordingly, whether minds are formal systems) without needless recourse to ontological aspects that, regardless of how technically right they might look, turn out to be ultimately irrelevant to the issue at hand. And, finally and ultimately, the case of Gödel's Theorem viewed as a formal model of the mind provides a crucial philosophical test case for Kant's doctrine of the synthetic *a priori*.

Viewed from this angle, Chomsky's thesis, i.e. his idea to specify all and only the (indefinitely many) expressions in a language, captured as a *formal system equal in force to that of the formalization of Peano arithmetic*, turns out to be of paramount importance, as it affords a necessary prerequisite and condition of a test for adequacy. In a strictly mathematical sense, Chomsky's thesis presents itself not as a "Gödel free" linguistic alternative to the formalization of mathematics, but as a mathematical system in its own right, wholly identical to the system of Peano arithmetic in all relevant respects (including Gödel's paradox); hence, equally alternative to all its higher-order formalizations, or else formalizations obtained by transforming undecidable theorems (so-called Gödel statements) into axioms. In the latter case, this leads to the construction of a larger theory in which the unprovable truths of the former theory assume the status of axioms. The latter move, however, does not yet make the new theory immune to Gödel's Theorem, and the proliferation of theories can go on this way indefinitely, no matter how many new axioms we latch on. In pure mathematics, the construction of a larger theory happens, however, to be only a purely hypothetical enterprise, since we must know in advance that this or that particular statement about numbers is true but unprovable. But the only way to know that this or that particular statement is true is to prove it. Since, however, this statement cannot be anything but unprovable, this means that we can never know that the statement in question is exactly the Gödel statement that needs to be added to the axiomatic base. Now, let us call this awkward property the Gödel predicament.

Since our axiomatic system is grounded in philosophical Rationalism and, therefore, it is itself a product of the theocentric paradigm in epistemology, consider that we are somehow in a position of obtaining God's view of things and thus we get free from the Gödel predicament as we find ourselves in a position where we know the Gödel statements that we need for the enrichment of the axiomatic base of our theory. Our new larger theory will then be nonetheless no more than only a restatement of the same problem, only in new terms, since one of the implications of Gödel's Theorem is that our system will always remain incomplete no matter how many new axioms we add. So all our new systems, no matter how large they get, will see yet new Gödel statements keep forever coming or growing like the new heads of Hydra's no matter how many we have already cut off. And, what is much worse, we will never know where they are coming from, why, and for what purpose... For this is not something that Gödel's Theorem itself can explain. Nor can this be seen. Not even looking from a God's eye point of view.

Things take a different turn, though, in the axiomatic formulation of *linguistics*. For ever since the time of Saussure's *Course in General Linguistics*, we no longer need to aspire to a God's eye point of view, since it was only in good old times before Saussure that linguistics was busy reconstructing God's genuine language, trying to penetrate to it through the raw facts of humble human languages remembering their original direct link to the divine essence of things, that had been dreadfully distorted in the wake of the events surrounding the construction of the Tower of Babel. The revolutionary step that Saussure accomplished was that he reshuffled accents, from historical linguistics with its glorious project of ascending to the divine roots of all languages, to synchronic linguistics regarded at that time as a perhaps curious yet not serious enough occupation for a theoretical linguist. Treating language synchronically, each language at any one time as a closed system in its own right without any reference to its past or future states, Saussure accomplished his own "Copernican turn" in general linguistics that parted with its former theocentric assumptions, becoming essentially and largely an anthropocentric discipline. Still later on, sampling his ideal speaker-hearer's "competence/performance" model on Saussure's opposition between *langue* and *parole*, Chomsky makes an anthropocentric view the backbone of his philosophy of language. Given Marr's known distinction between three levels of explanation of a computational system, distinguishing between the top level of computational theory, the middle level of representation and algorithm, and the bottom level of hardware implementation, and his claim that we understand an information-processing system completely only when we understand it at all three levels,[19] it is only natural that Chomsky foresees *bio*linguistics

[19] D. Marr, *Vision: A computational investigation into the human representation and processing of visual information*, p. 24.

as exactly that what should be posited at a respective bottom explanatory level as a linguistic counterpart playing the role of hardware in Marr's theory of computational system as a wanted and unavoidable complement to linguistic science, Postal's objection to biolinguistics as Chomsky's ontological position quite notwithstanding.[20] Hence, since the production of new utterances turns out to be a self-governed and self-controlled human linguistic performance within the bounds of the samely human linguistic competence, we can reasonably predict, that the so far unattainable ideal of a God's-eye view of things may acquire somewhat more human dimensions and thus, in a sense, become more real.

On the face of it, however, Collins is definitely wrong about the status and interpretation of Gödel's results in claiming on behalf of Chomsky that they have no particular bearing on linguistics.[21] All consistent axiomatic formulations include undecidable propositions. No axiomatic formulation is an exception to this rule. In generative-transformational linguistics likewise, all formal properties of the mathematical empire still ensue no matter what, Emperor's *new clothes notwithstanding*. Evidence to this effect (due to a vast proliferation of theories) abounds throughout the history of formal linguistics (vide the history of Extended Standard Theory (EST), Revised Extended Standard Theory (REST), Government-Binding (GB) theory, the Minimalist program), suffice it here to mention the introduction, some quarter of a century ago, of a new underlying level of linguistic description, that of Logical Form (abbreviated LF), as an attempt to resolve the problem of the syntactic derivation of multiply quantified sentences, which gives a clear example of a construction of a larger theory by amending the axiomatic component of a formal system.[22] This innovation was aimed to provide a systematic account of semantic deviations accompanying the otherwise meaning-preserving transformations, when the latter were applied to quantifier structures, explained in terms of their structural descriptions at LF, the syntactic level fully disambiguated with regard to their logically relevant semantic properties, and thus represented as part of their overall syntactic analysis. As May points out, "ambiguities of multiple quantification are therefore syntactic ambiguities, grammatically disambiguated, a "constructional homonymity."[23]

[20] Paul Postal, *The Incoherence of Chomsky's 'Biolinguistic' Ontology*, p. 104.

[21] John Collins, *A Question of Irresponsibility: Postal, Chomsky, and Gödel*, pp. 102–3.

[22] William A. Ladusaw, *Logical form and conditions on grammaticality*; R. May, *Logical Form as a Level of Linguistic Representation*.

[23] R. May, *Logical Form. Its Structure and Derivation*, p. 16.

Yet further analysis would nonetheless inevitably detect the existence of "constructional homonymity" at LF itself, and thus the necessity of assuming yet another level of syntactic disambiguation, this time with respect to the syntactic representations at LF, in addition to, and along with, the syntactic disambiguation provided by LF with respect to S-Structure.[24] Since both levels would be specified with regard to one type of interpretation only, yet further ambiguities that come inevitably at the upper level of LF would as well result in the need to explain them by assuming still further levels of LF, this eventually leading to infinite regress, fully in accordance with the implications of Gödel's Incompleteness Theorem.

There are, therefore, at least two reasons for which the above-mentioned introduction of LF as an additional level of linguistic explanation is worth noting here. In the first place, it provides a quite seizable example of the ongoing construction of larger theories, rather than a mere conception of it as a hypothetical implication of Gödel's Theorem, as is the case in the axiomatic formulation of mathematics. And, what is most important for the point at issue, it also provides a sound demonstration of a more powerful epistemological perspective that arises in the linguistic counterpart of the axiomatic formulation of Peano's arithmetic. That is to say, instead of possessing the property of being "Gödel free" that Chomsky assumed his axiomatic formulation of linguistics has, it turns out to be entertaining the property of being free from Gödel predicament. For in the latter case we seem to happen to be obtaining the crucial and immediate knowledge of the whereabouts of Gödel statements and the ability to identify them as such, that was otherwise denied in the humanly unattainable theocentric perspective of a God's-eye point of view.

The immediate knowledge in question is knowledge about whether this or that previously given statement is true but undecidable. If a given statement is about natural numbers, we do not happen to be in a position to produce a concrete example of such a statement (thus this statement exists only by hypothesis), because in the first place we would have to somehow decide that such a statement is true, which we cannot do as it is undecidable. We could of course know such things intuitively but, in mathematical matters, intuition may not help. What is evident to one person may be not so evident to another. Even in the case of the so-called evident axioms it takes a certain amount of mathematical sophistication to appreciate. And even though, as Gödel suggested, in his famous article

[24] *Ibid.*, p. 15.

on Cantor's continuum problem, that what is evident can be cultured, it is not at all evident what makes this or that axiom evident.[25] Leaving the question of what makes the axiom evident aside for the moment, let us take for granted Gödel's suggestion that it takes a certain amount of mathematical sophistication to be able to judge whether this or that axiom is true, and that the level of sophistication can be cultured and thus is an element of overall mathematical culture, it is easy to see that the respective ability to judge the properties of linguistic forms is an analogous element of the respective linguistic culture; moreover, it follows directly from the very fact that the person in question is a native speaker of the language of this culture. Thus, we can judge whether one sentence in English means the same thing as another sentence of English, or whether a linguistic expression in that language has more than one meaning. We know, for example, that such is the phrase *the shooting of the hunters*. The knowledge that *the shooting of the hunters* has two different meanings, and that it does not depend on the meanings of its lexical constituents, is a linguistic truth, a theorem about language that we know is true. The second question is how do we know that it is true. According to Chomsky's claim, to answer this question we show how it is derived from two different axioms in the axiomatic formulation of a language provided by transformational-generative grammar strictly in line with Chomsky's thesis.

However, this claim runs into immediate problems when we come across a multiply quantified sentence like *everyone loves someone* or *every boy danced with a girl*. That either of these forms has two different meanings is a fact instantly evident to every native speaker or a person competent in this or that language in the case of their equivalents in German, Spanish etc. Then we reach for transformational-generative grammar for an explanation of why it is so that this case of "constructional homonymity" is as evident as it is evident in the similar case with *the shooting of the hunters* that we know is produced or derived from two different underlying syntactic structures, *someone shot the hunters* and *the hunters shot someone*. The standard version of transformational-generative grammar will not do to explain this case, however, as its axiomatic base will not be sufficient for providing such a proof. Hence, what we are dealing now at this stage is a theorem that is true but unprovable. In other words, the very fact that in our system the "constructional homonymity" in question is undecidable provides clear evidence that it is nothing other than a

[25] S. Feferman, *Are There Absolutely Unsolvable Problems? Gödel's Dichotomy*, p. 143.

language specific, particular case of Gödel statements. Henceforth, what we are bound to do next is exactly what Gödel's Theorem implies, that is, we have to modify the axiomatic component of the theory so as to make our Gödel statements decidable by adding new lines of proof from the corresponding new axioms, expanding the theory by the necessary implementations, and naming the so enforced additional explanatory level as 'a new head of Hydra', or 'a new level of LF', or whatever we choose.

Suppose now, that we have convinced ourselves that our extended theory supplies the required explanation by postulating the existence of an additional explanatory syntactic level of Logical Form, at which the difference between the two distinct meanings of *every boy danced with a girl* (one implying the situation in which each boy danced only with the girl he came with, yet another, that a certain girl danced in turn with each of the boys who came to the party) is represented as a difference between the respective syntactic structures at the level of LF, either of which having its own derivational history (theorem's proof):

(3.1) a. $[_S [_S$ every boy$_x$ $[_S$ girl$_y$ $[_S$ x danced with y]]]]
b. $[_S [_S$ girl$_y$ $[_S$ every boy$_x$ $[_S$ x danced with y]]]]

The meaning difference is then explained by the fact that at the level of LF the quantifier occupies different structural positions and thus has different scope (which it does not do at the level of S-Structure). Since LF is a level of logical form, it is also assumed that "such disambiguation that we find at LF, under our syntactic characterization of this level, will clearly be relevant in determining logical consequence in natural language (with respect to a specified semantic interpretation).[26] On this assumption, the relation of logical consequence ought to be exactly the relation between the two variant meanings of *every boy danced with a girl.* And, hence, it happens to be the relation between (3.1b) and (3.1a), which does not show itself directly at the level of LF, but can nonetheless be demonstrated by applying the notation of first-order predicate calculus:

(3.2) $\exists y \forall x F(x, y) \rightarrow \forall x \exists y F(x, y)$

This logical notation, used as a means to capture the relation of logical consequence between (3.1b) and (3.1a), spells out that if $\exists y \forall x F(x, y)$

[26] R. May, *Logical Form. Its Structure and Derivation*, p. 16.

(where the universal quantifier has a narrow scope) happens to be true, then $\forall x \exists y F(x, y)$ (where the universal quantifier has a wide scope) is, by logical necessity, also true. The recognition of this logical fact does not, however, imply that if there was a certain girl that every boy danced with, then, necessarily, every boy should have also danced with the girl in whose company he came to the party. We would be right in saying that *every boy danced with a girl* implies (*in the sense of* 'presupposes' or 'assumes') either that (a) every boy danced each with the girl he came to the party or that (b) every boy danced with one and the same girl, but it is equally evident that neither (a) implies (b), nor (b) implies (a), for these are quite separate and logically independent facts. Yet, since, at LF, we use the difference in the scope of quantifiers as a means for the explanation of the difference in meaning between (a) and (b), we cannot ignore the fact that the essential property of quantifier scope is that the broader scope implies the narrower one; hence, the relation of logical consequence from the narrower scope to the broader one, as shown symbolically in (3.2), as well as represented graphically in (3.3) below:

(3.3)

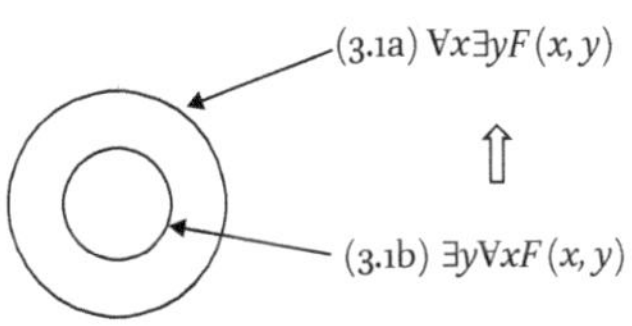

The corollary of the fact that the logical explanation employed at LF shows (a) and (b) as logically connected by the relation of logical consequence (3.3), though they are, in fact, logically independent as their meanings are quite disjunct, as shown in (3.4):

(3.4)

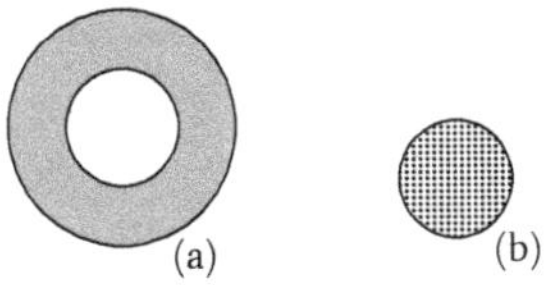

is that the expression (3.1a) at the level of LF turns out ambiguous between the (a) and (b) interpretations exactly in the same way as the sentence *every boy danced with a girl* is ambiguous between the meanings of (3.1a) and (3.1b) itself:

(3.5)

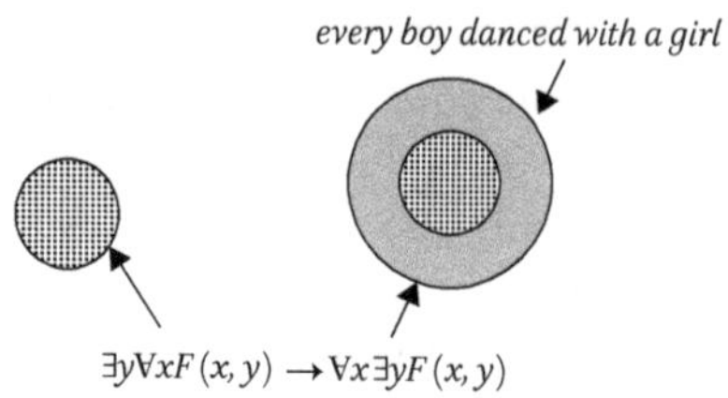

Since the situations (a) and (b) are distinct, logically independent facts, then, contrary to May's claim, there can be no relation of logical consequence between (a) and (b) at all. What in fact makes the relation of logical consequence in $\exists y \forall x F(x, y) \rightarrow \forall x \exists y F(x, y)$ at all possible is the very fact that the consequent, $\forall x \exists y F(x, y)$, is itself ambiguous between the interpretations (a) and (b) in (3.4), and this is the only reason for its being true if the antecedent, $\exists y \forall x F(x, y)$, is true [see the right side of the diagram in (3.5) as compared to the left side of the respective diagram in (3.4) where the meanings of (a) and (b) are shown as disjoint]. By the same token, knowing that "at least two languages are known by everyone in the room" is true, we can also infer that "everyone in the room knows at least two languages" is also true. All this goes to show that the "constructional homonymity" found at the level of S-Structure in, say, "every boy danced with a girl", repeats itself as an inevitable "constructional homonymity" of the respective logical structure in "$\forall x \exists y F(x, y)$" and, by the same token, of the analogous structure in "$[_S [_S$ every boy$_x$ $[_S$ girl$_y$ $[_S$ x danced with y]]]]" at the level of LF itself. Thus, what it turned out that had been achieved by introducing the additional explanatory level of linguistic explanation of LF was, to use Hornstein's turn of phrase, "not so much an explanation as a repetition of the problem"[27] and, thus, a clear instance of the corollary of Gödel's Incompleteness Theorem, namely that a formal system is doomed to remain incomplete, no matter what. Not even if it appears in the disguise of a linguistic theory.

5.4. *Are There Absolutely Unsolvable Problems? Gödel's Dilemma*

Let us now consider if the change of epistemological perspective that we have been elaborating on so far has any impact on an ontological issue,

[27] N. Hornstein, *Logic as Grammar. An Approach to Meaning in Natural Language*, p. 122.

raised by Gödel, regarding the existence of absolutely unsolvable problems. As Gödel himself explains,

> [T]he epithet "absolutely" means that they would be undecidable, not just within some particular axiomatic system, but by *any* mathematical proof the mind can conceive.[28]

Gödel dwells upon this ontological problem in the basically epistemological context of what Feferman in his paper on Gödel calls "Gödel's dichotomy" where it appears as one of the alternatives in a disjunction:

> Either mathematics is incompletable in this sense, that its evident axioms can never be comprised in a finite rule, that is to say, the human mind (even in the realm of pure mathematics) infinitely surpasses the powers of any finite machine, or else *there exist absolutely unsolvable diophantine problems*...[29]

By a *diophantine problem* is meant a proposition of elementary number theory of a relatively simple arithmetical form whose truth or falsity is to be determined; its exact description is not important to us.[30] On Chomsky's thesis, since a finite machine in question is "the grammar of L [which is] a device that generates all of the grammatical sequences of L and none of the ungrammatical ones,"[31] a *diophantine problem* is, accordingly, an expression in L whose grammaticality or ungrammaticality is to be determined. And, "to say that the human mind is equivalent to a finite machine "even within the realm of pure mathematics" is another way of saying that what the human mind can in principle demonstrate in mathematics is the same as the set of theorems of some formal system."[32] Notice, that what is at issue in Gödel's question is really not an ontological problem regarding the existence of certain mathematical truths, but an epistemological problem about the epistemic limits of knowledge. The real question turns out therefore to be not about truth as such, but about matters that make our knowledge of mathematical truth possible. For, to say that the human mind is equivalent to a finite machine is to say that the theory of human mind is equivalent to that of a formal system. And, saying that the human mind surpasses the powers of any finite machine amounts to saying that

[28] Kurt Gödel, *Collected Works, Vol. III: Unpublished Essays and Lectures*, p. 310.
[29] *Ibid.*
[30] S. Feferman. *Gödel, Nagel, minds and machines*. Ernest Nagel Lecture, Columbia University, Sept. 27, 2007.
[31] N. Chomsky, *Syntactic Structures*, p. 13.
[32] S. Feferman. *Gödel, Nagel, minds and machines.*

the human mind has a mathematical insight not derivable from the axioms of some formal system.[33]

On Feferman's analysis of Gödel's dilemma, the problem of whether absolutely unsolvable problems exist remains, however, unresolved. For, as Feferman concludes:

> We do not have any precise criterion for the solvability of individual problems which would allow us to prove the existence of problems that are absolutely unsolvable in principle; so it is idle to ask for examples of such.[34]

What he points at, instead, are the two problems that in his opinion are only candidates to be absolutely unsolvable: Cantor's continuum problem, and problem (P1) below:

> (P1) Is the value of the digit in the $10^{10^{10^{10}}}$th place of the decimal expansion of $\pi - 3$ equal to 0?

About (P1) he says that

> this is an example of a mathematical 'yes/no' question, whose answer can be determined in principle by a mechanical check, but which, in all probability, cannot be settled by the human mind because it is beyond all remotely conceivable computational power on the one hand and there is no conceptual foothold to settle it by a proof on the other.[35]

In a way, the problem of the syntactic derivation of multiply quantified sentences discussed above presents itself as a close analogue of Cantor's continuum hypothesis. The latter assumes that if $X \subseteq R$ is an uncountable set, then there exists a bijection $\pi : X \to R$. A bijection is a one-to-one correspondence between elements of two sets. The hypothesis of LF, as a theory of multiple quantifier structures and their syntactic derivation, is exactly a theory of bijection between the different syntactic structures derived at the level of LF (as members of set R) on the one hand, and the different meanings of multiply quantified sentences at the level of S-Structure (as members of set X) on the other, such that either of the underlying structures at LF [see (3.1a)–(3.1b) above] corresponds to exactly one of the respective meanings [(a) or (b)] of an expression at the level of S-Structure (such as *every boy danced with a girl*). And, as an

---

[33] Kurt Gödel, *Collected Works, Vol. III: Unpublished Essays and Lectures*, p. 309.

[34] S. Feferman, *Are There Absolutely Unsolvable Problems? Gödel's Dichotomy*, p. 149.

[35] *Ibid.*, pp. 149–50.

example of a mathematical 'yes/no' question, whose answer is assumed to be determined in principle by a mechanical check, this problem, in all probability, cannot be settled by a finite machine because it is beyond all remotely conceivable computational power exactly for the reason that there is no conceptual foothold to suggest any optimism [see diagrams (3.3)–(3.5) above] as to whether at some computational stage of determining the existence of a bijection between multiply quantified structures of different levels we might have arrived at a "yes" answer.

An even more evident example of the existence of problems that are absolutely unsolvable, this time more closely to the kind of Feferman's problem (P1), is rendered by the structurally more simple cases of non-multiple (phrasal) quantifier constructions that need not involve any "constructional homonymity" that needs to be treated by bogus explanatory structures at LF with no way to assuage lingering set-theoretic worries about it. Since, due to its relative simplicity, the phrasal syntax of quantifiers inevitably falls out of the descriptive and explanatory purview of the logical analysis of quantifiers as variable-binding operators for the same reason for which it necessarily falls out of the specific domain of polyadic predication and multiple quantifiers, all relevant problems connected with these syntactic constructions amount to a simple yes or no question regarding their grammaticality. Since in this case a diophantine problem is not a mathematical judgment, but a grammatical one, the relevant question about a Gödel statement whether it is true that σ (Gödel statement) is true is rephrased, accordingly, as a yes or no question whether it is true that σ is grammatically correct. In more precise terms, the question about the decidability of σ is the one of asking if it is true that σ (as a number-theoretic statement) is either true or false or, respectively, if it is true that σ (as a linguistic expression) is either grammatically correct or ungrammatical.

The latter is a direct consequence of Chomsky's thesis and, by its importance, it stands on a par with the invention of Cantor's diagonal argument, or of a Gödel numbering. The latter two are lucky exceptions, rather, since what makes a really great thinker is exactly that one can rarely see (and foresee) all the consequences and the real importance of his own results. Of course, Cantor and Gödel invented their tools as part of their general strategy, whereas Chomsky's concern was specifically linguistic rather than a foundational one. This is exactly why he failed to see the paramount importance of his own invention as a kind of diagonalization procedure to provide an escape route from Gödel's paradox of "true but unprovable" statements in the same way as in Gödel's own case was accomplished by Gödel numbering so as to break the circularity of self-reference of a lier paradox in "This statement is false" (a case seems to be similar to that of Löwenheim's theorem in his paper, „Über Möglichkeiten im Relativkalkül" (1915), that presented the first significant result in what later became model theory, the fundamental importance of which Löwenheim did not seem to recognize, as his interest in the first-order fragment he had demarcated seems motivated by purely algebraic, rather than foundational considerations).[36]

Chomsky's own position on the issue seems, however, to be that of the constructive mathematician, since his assumption that in linguistics, "no 'Platonism' is introduced, and no 'E-linguistic' ('external' and 'extensional') notions: only biological entities and their properties,"[37] can be easily traced down to the strictly constructivist interpretation of the phrase "there exists" as "we can construct." In Brouwer's philosophy, known as intuitionism, mathematics is a free creation of the human mind, and an object exists if and only if it can be (mentally) constructed.[38] There is, therefore, an observable parallelism between Chomsky's conviction that, in his axiomatic formulation of language, there are no (or, rather, there cannot be any) undecidable propositions, and a constructivist proof, by Per Martin-Löf, that "There are no propositions which can neither be known to be true nor to be known to be false."[39] And, indeed, it seems trivial to see that proofs that show one how to construct the object being proved, cannot reveal the existence of objects that these proofs do not show how to construct. As Feferman rightly concludes:

> For the non-constructive mathematician, Martin-Löf's result would be translated roughly as: No propositions can be produced of which it can be shown that they can neither be proved constructively nor disproved constructively. For the non-constructivist this would seem to leave open the possibility that there are absolutely unsolvable problems A 'out there', but we cannot produce ones of which we can show that they are unsolvable.[40]

---

[36] W. Goldfarb, *Logic in the Twenties: the Nature of the Quantifier*, p. 355. Yet, as Goldfarb notes, "it was just this algebraic background that enabled Löwenheim to arrive at the remarkable result and at the deep idea needed for the proof" (*Ibid.*).

[37] N. Chomsky, *Derivation by Phase*, p. 42.

[38] Stanford Encyclopedia of Philosophy: Constructive Mathematics.

[39] Per Martin-Löf, *Verificationism Then and Now*, p. 195.

[40] S. Feferman, *Are There Absolutely Unsolvable Problems? Gödel's Dichotomy*, p. 147.

Despite all appearances, however, the application of the axiomatic method in linguistic theory, accomplished by Chomsky, provides an excellent opportunity to perform the kind of diagonalization that would be indispensable for solving the problem of Gödel's "true but unprovable" statements. As already mentioned, since in the case of generative-transformational grammar a diophantine problem is not a number-theoretic judgement, but a grammatical one, the relevant question about a Gödel statement whether it is true that σ (Gödel statement) is true needs, then, to be put another way, as a yes or no question whether it is true that σ is grammatically correct. Chomsky's own "constructivist" argument then stands or falls on our ability to produce the linguistic counterparts of absolutely unsolvable problems 'out there', of which we can show that they are unsolvable. And, even more than that, if these problems can be shown as intimately intertwined with the deficiency of our understanding of quantity, we will also be able to diagnose the exact cause of their being "absolutely unsolvable" by tracking the source of the problems, down to the deficiency of the notion of quantity. Ultimately, to ask for examples of such "absolutely unsolvable" problems turns out to be not as idle as Feferman expects, exactly due to the fact that in this case we obtain a precise criterion for the solvability of individual problems.

What Feferman has in mind speaking of the criterion in question:

> We do not have any precise criterion for the solvability of individual problems which would allow us to prove the existence of problems that are absolutely unsolvable in principle; so it is idle to ask for examples of such.[41]

is not, of course, that we need a criterion which, as he puts it, "would allow us to prove the existence of problems that are absolutely unsolvable in principle," for this task has been already accomplished by Gödel himself, and the criterion in question is already contained in the proof of his incompleteness theorem. The attribute "individual" is, of course, the key word here, and what is to be decided is not the hypothetical existence of unsolvable problems *in toto*, but the real existence of this or that problem that we could deictically point to as existing "out there" so as to produce a 'living' example of the so far only hypothetical existence. Besides, even if Feferman does not make it clear, it is evident that the criterion he is speaking about, is not intended for showing that this or that particular

[41] S. Feferman, *Are There Absolutely Unsolvable Problems? Gödel's Dichotomy*, p. 149.

problem is solvable; rather, what he has in mind is a precise criterion for the *un*solvability of this or that individual problem.

5.5. *Gödel's Dichotomy: The Third Alternative*

In order to point to the real problems "out there" and to look for such a criterion, let us consider the so far failed attempts to construct a satisfactory theory of the transformational derivation of quantifier expressions. As the "abnormal" semantic properties of quantificational structures were first noted, by Chomsky, at the very start of the construction of an axiomatic system in linguistics (his branded examples relate to the non-synonymity of "everyone in the room knows at least two languages" and "at least two languages are known by everyone in the room"), the suspicion arose that quantifier items in English need to be given special treatment due to the manifestly logical character of their meaning. This seemed to have been confirmed by the close affinity between the reduction of quantifiers to conjunctions or disjunctions of singular predications in logic and the respective behaviour of quantifiers and conjunctions in English, which led to the rise of a paralogical linguistic conception of NL-quantifiers as higher predicates. This conception, as was first noted by Guy Carden back in 1967, proved however useless insofar as the syntactic status and derivation of prepositional quantifier structures were concerned. Yet another attempt to explain the syntactic derivation of phrasal quantifier expressions, based on the ideas of set theory, was undertaken by D. A. Lee back in 1971, however unsuccessfully. The so far most in-depth study addressing various issues of the transformational derivation of phrasal quantifier expressions dates back to Jackendoff's 1968 theory, essentially unaccomplished. What this effectively means is that while the meaning of the linguistic expressions with quantifiers is pretty well understood and their grammaticality is perfectly evident (the latter being the condition of the former), we are yet lacking an explanation of *why* we do understand what we understand, so what is critically lacking is an account for their grammaticality in terms of showing the analytical transformational mechanism of their derivation. In effect, we are facing a special case of the Kantian problem of understanding. Being translated into the vernacular of the theory of formal systems, this means that what we are dealing with is a theorem that we know is true, while the fact that we have no proof of this theorem may mean either that no such proof is known *at this point* or else that no such proof is possible, the latter being the case of Gödel's true but unprovable theorem (Gödel's "absolutely unsolvable problem"). In light of the analysis undertaken earlier in this study, it is perfectly clear that no such proof is in deed possible, even in principle. In the same way and for exactly the same reason, trying to find such a proof is as vain an enterprise as trying to find a pronominal form for indefinite mass nouns in English, of which it is perfectly known that it does not exist. From the point of view of language as a formal axiomatic system, then, the syntactic derivation of quantifier expressions in question presents itself as exactly the kind of Gödel's absolutely unsolvable problem that we can point to "out there."

Since quantifier expressions are thus instances of "evident axioms [that] can never be comprised in a finite rule," with the latter being equal to the statement to the effect that there exist absolutely unsolvable problems that we can point to out there, it will be safe to say that, in the long run, both disjuncts in Gödel's dichotomy turn out to be true. This is exactly what Gödel himself asserts in a parenthetical remark directly following the statement that 'the case that both terms of the disjunction are true is not excluded, so that there are, strictly speaking, three alternatives'.[42] And, considering the philosophical consequences of his dichotomy, he concludes that under either alternative they support "some form or other of Platonism or "realism" as to the mathematical objects."[43] Or, to put it in other words, no matter whether the first term of the disjunction is asserted and the second one is negated, or the first term of the disjunction is negated and the second one is asserted, either case equally implies transcendental realism and the analytic *a priori* as the philosophical bases of mathematics. Things turn out to appear very differently with the third alternative, however. For the philosophical consequences in the case of the third alternative lead to a total negation of both analytic *a priori* and transcendental realism alike. And in no way do they happen to support "some form or other of Platonism or "realism" as to the mathematical objects," for the attainment of God's eye point of view, from which we can see *all* true Gödel statements as true without getting any chance of knowing *why* and *how* we can see them as true, leads us to having no real meaning for truth. In a way, we find ourselves confronted by a kind of the Tower of Babel curse rolling back on us for the sin of aspiring to be God.

---

[42] *Ibid.*, p. 148.

[43] Kurt Gödel, *Some basic theorems on the foundations of mathematics and their implications*, p. 311.

As a corollary, what this effectively means is that we are confronted with two options: either that 1. Gödel statements are exceptions to the rule and are thus exceptions to analytic *a priori*; hence, synthetic *a priori*; or that 2. the existence of Gödel statements is necessary; thus, all mathematical truth *in toto* is necessarily synthetic *a priori*, with analytic *a priori* being only the bordering, or limiting case of synthetic *a priori*; in short, being itself only a kind of synthetic *a prori*.[44]

Let us now proceed considering these options step by step, starting with the consideration of the weaker claim, namely that quantifier expressions in English as unprovable theorems in an axiomatic formulation of linguistics are exceptions to analytic *a priori*; hence, the only option left for them is to be synthetic *a priori*. The truth of the statement that in an axiomatic formulation of linguistics these expressions are not derivable from an analytic basis follows directly from the analysis of two mutually opposing generative schemata, those of "*ones*-absorption" and "*ones*-extraction":

(5.1) *Ones*-absorption:

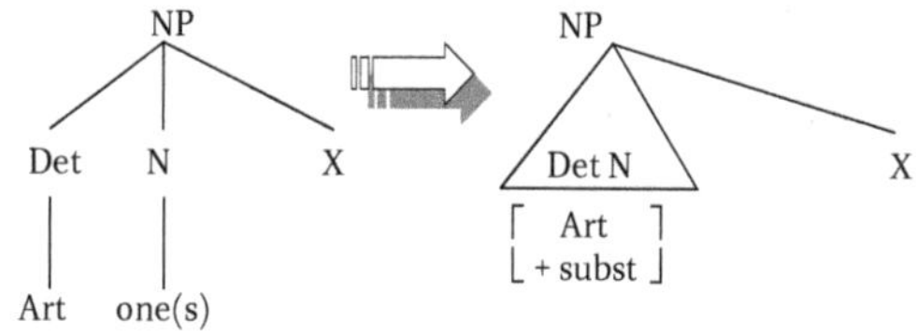

(5.2) *Ones*-extraction:

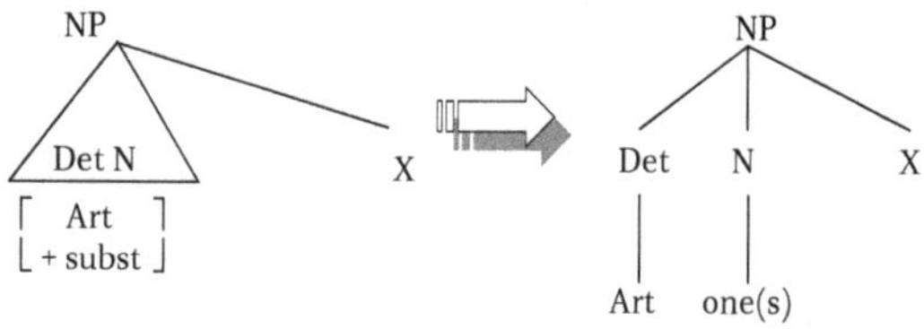

Moreover, as has been already demonstrated in Chapter 4, the generative scheme in (5.2) turns out to be the only explanatory pattern that effectively describes all occurrences of quantifier expressions, both with and without count nouns. In contradistinction to (5.1) in which the derivation

[44] ... the latter being, in fact, a complete reversal of Frege's view on synthetic as "kind of analytic." Cf. discussion on Vendler's "lawlike propositions" above, pp. 154–55.

starts from an initial tree of an *a priori* analytic base component of generative-transformational grammar, in (5.2) all derivation proceeds from a manifestly synthetic base. Furthermore, conflating these two schemata in one general *extraction* vs. *absorption* scheme:

(5.3) Ones-e*xtraction* vs. ones-*absorption*

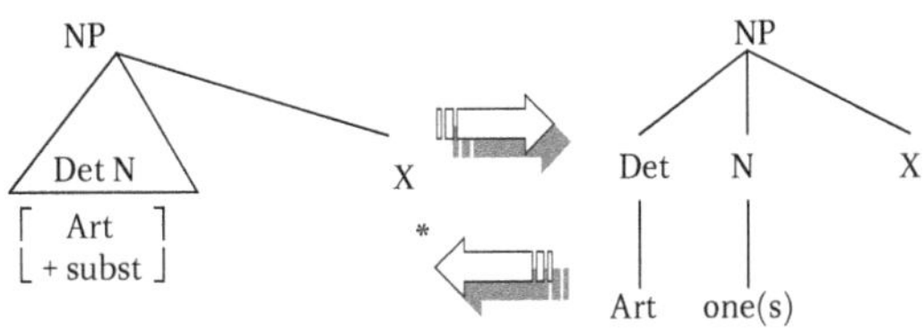

effectively demonstrates that the relation is irreversible and, thus, has a clear "cause-effect" status. In virtue of this, the scheme that accounts for the grammaticality of quantifier expressions also accounts for the synthetic *a priori* as a genuine source of their grammaticality. Keeping in mind that in Chomsky's axiomatic formulation "grammaticality" is nothing but only a linguistically disguised notion of "truth" that we find in an axiomatic formulation of mathematics,[45] this also shows that the analytically defined "formal truth" turns out to crucially depend on Kant's "material truth" as a proper subject of transcendental logic and, thus, on an *a priori* synthetic base as its necessary condition.

Now, the question arises if what we are dealing with happens to be merely a local phenomenon that has no effect on whatever that resides outside the boundaries of linguistic theory proper. Reasoning this way, we would be exposed to the same sort of Chomsky's "minds are not machines" fallacy, for this would require positing an assumption that Chomsky's axiomatic formulation rests on something other than a theory of formal system, which is manifestly wrong. Secondly, we would have to admit that Gödel's Incompleteness Theorem is not universally valid and, therefore, is not a theorem. But this is impossible as we would then need to prove, no more and no less, that what is not valid is Gödel's proof as such. And, ultimately, we would have to claim that the meaning of quantitative expressions in a language is entirely different from the meaning of

[45] ... which is exactly the point of Chomsky's "diagonalization".

those in arithmetic, and that, in the long run, number expressions stand for something other than numbers.

Thus, if we accept the view that Chomsky's linguistics is simply a particular case of the theory of formal systems (and nowadays one can hardly argue otherwise), it would be hard to deny that "Chomsky's thesis" (Bach's term) turns out as a genuine diagonalization method that enables the demonstration of the ultimate solution as to the (so far heatedly debated) philosophical consequences of Gödel's Incompleteness Theorem. Moreover, the solution in question bears directly on the most fundamental issue in the philosophy of mathematics, that of the nature of numbers. Since, as has been stated above, numbers are not objects, let us consider briefly the alternative to "some form or other of Platonism or realism" that this latter solution provides in terms of its own philosophical consequences. For this happens to be exactly the alternative assumed by Frege's question:

> So that a sentence like 'four is a square number' does not express the subsumption of an object under a concept. And so this sentence cannot at all be construed like 'Sirius is a fixed star'. But then, *how*?[46]

It will be seen that the earlier alternative, that Frege comes to reject in a later entry to his diary in March 1924, goes back to his 1884 *Grundlagen* theory, according to which both noun phrases like *the number of planets* and simple numerals like *eight* as in (5.4) are singular terms referring to numbers as abstract objects:

(5.4) The number of planets is eight.[47]

Hence, as Frege claimed (against Kant), "nought and one are *objects* which cannot be given to us in sensation."[48] And this was exactly the claim that Gödel thought the philosophical consequences of his Incompleteness Theorem supported in the case of either of the alternatives in Gödel's dichotomy being true. But the truth of the third alternative (both of the former being true) provides clear evidence against this earlier claim of Frege's; moreover, it also provides a clear solution to Frege's unanswered question. Specifically, as regards Frege's claim that "nought and one are objects which cannot be given to us in sensation," it does support Frege's contention that *one* cannot be given to us in sensation. But the reason for

---

[46] G. Frege, *Nachgelassene Schriften*, p. 282.
[47] F. Moltmann, *The Number of Planets: A Number-Referring term?*
[48] G. Frege, *The Foundations of Arithmetic*, p. 101.

which it cannot be given to us in sensation is entirely different from that of *one* being an object of a Platonic kind, as Frege had originally thought. For the real reason turns out to be wholly vindicating Kant's position on the issue. Namely, the reason for which *one* cannot be given to us in sensation is that it relates to sensation in exactly the same way in which apperception relates to perception. Specifically, *one* turns out to be an *a priori* synthetic conception that forms, in Kant's sense, the ground of reference to objectivity. For, as the foregoing analysis has shown, *one* has, as necessary elements of its meaning, a dualistic pair of *a priori* conceptions of quantity with a unique characteristic of non-homogeneity, that of "discreteness vs. continuity." Note that the latter conceptions are obviously *a priori*, as they are clearly not following from, nor representing the inner quality of the stuff/objects conceived, but only the *a priori* preconceptions, since "a sack of sand" or "a sack of coffee" from a plantation will be equally preconceived as sacks of indivisible stuff quite regardless of the fact that the stuff in question is, objectively as well as perceptually, a collection of quite evidently discrete elements, i.e. grains of sand and coffee beans. A herd of cattle or a stock of geese added to another herd of cattle or stock of geese makes even though twice as large yet not two but still one herd or stock, so in this case a simple arithmetical operation "1 + 1 = 1" will yield *one*, not *two*. Which shows that the meaning of *one* is a matter of organizing our perceptions, of individualizing their material substrate, of betokening a type,[49] thus, a necessary precondition of our recognizing objects as such, making the very ground of reference to objectivity, the cardinal question about the nature of which was posed by Kant as the main issue of transcendental logic. Therefore, contrary to Frege and Gödel (and the whole of analytic tradition), neither numbers nor the laws governing them can have any locus in objectivity, not even in its Platonic version. So, to paraphrase Albert Einstein's famous quote, "To the extent math refers to reality, we are not certain; to the extent we are certain, math does not refer to Platonic reality." For its genuinely primary locus lies not in objecthood, but in the very *conditions* of objecthood *per se* (relative to the pure intuitions of space and time); thus, numbers find their locus in the necessary conditions of any discursive thought whatsoever, which, at bottom, ultimately makes Gödel's much expected case of "Kant understood correctly." In effect, Frege's conception of mathematics as an extension of logic turns out to be true in exactly the opposite sense,

---

49 Henry E. Allison, *Kant's Transcendental Idealism*, p. 118.

with the respective notions taken in the reverse order. For the present case study of the conception of quantity in the framework of transcendental logic provides ready evidence directly to the opposite effect; namely, that it is mathematics, in its *a priori* synthetic foundations, that lies in the very basis of all discursive thought.

CONCLUSION

The present volume has been concerned with both the philosophical significance of Kant's problem of *a priori synthesis* as well as the genuine mechanism of this synthesis, the study of which Kant contemplated as a prime task of transcendental logic viewed by him as a prospective autonomous academic discipline, differing from what he viewed as "general logic" in both its assignment and method. The rationale for the present study of the genuine mechanism of *a priori synthesis* is as follows: (1) Transcendental logic as it was envisaged by Kant cannot exist only in blueprint form: yet further work on a construction site is simply a logical imperative for the enterprise to make sense at all; (2) in order to do full justice to the core of Kant's thought, there is a pressing need to upgrade the study of Kant, from a mere elucidation of the literal meaning of Kant's words through the in-depth discernment of Kant's views against the philosophical background that he was reacting to, to a basically new level that, as Gödel put it, transcends "common sense" and thus provides grounds for a better understanding of what Gödel believed, in his words, "to be a general feature of many of Kant's assertions that literally understood they are false but in a broader sense contain deep truths."

The latter task requires putting Kant's critical philosophy in a much broader context that would thus necessarily involve a critical reassessment of the developments of philosophy after Kant that bear crucial, critical or otherwise, relation to the basic elements in Kant's train of thought. Since the key distinction in the Kantian philosophy is the distinction between Reason and Sensibility, with Kant's central issue being that of the nature of synthetic *a priori* with the main emphasis being put on an essential cognitive role of Sensibility, such basic element appears to be that of "betokening a type" and thus revolves around the opposition between generality and particularity, both of the latter comprised under the traditional notion of "logical quantity." The first part of the present study was thus devoted to a critical reconsideration of Frege's work as a founding father of the present-day, essentially anti-Kantian, analytic tradition in philosophy so as to trace the crucial elements of the evolution of philosophical understanding, after Kant, of the relevant notion of "logical quantity" disguised in a notion of the quantifier as the basic element of modern "quantificational logic."

A closer look at the earliest stage in the development of analytic tradition shows that Frege came very close to Kant's idea of "synthesis," as his concavity with the Gothic letter inscribed (Frege's original expression of "generality") served not only as a divide, but also as a connector between the general and the particular logical content of the same expression, showing them as the two sides of one and the same coin. There was yet another deeply philosophical touch in Frege's innovation, that looks quite Kantian in spirit, for Frege's purpose behind the addition of the stroke standing for general logical content was none other but an attempt of an *a priori* imposition of a general lawlike nature on the particular logical content of Frege's conditional that in his view expressed a cause-effect relationship between the two saturated concepts as parts of a categorical statement, thus, just because of their being saturated by specific argument values, necessarily depicting the particular content of the conditional, therefore, only placeable to the right of Frege's divide between the universal and the particular. This Kantian "undertone" of Frege's thought was, however, neglected or else totally misunderstood in post-Fregean tradition, in which the logical quantifiers, specifically in Quine's line of thought, came to be seen as bearing no deeper meaning than simply that of the logical "quantificational" idioms in "for all *x* such that..." and "there is at least one *x* such that..."

The undertaken analysis shows that the reason of misunderstanding Frege lies in his purely technical exposition of his *Begriffsschrift*, meant as a technical instruction of how to use it when applied in the workshop of a mathematician. Much of Frege's talk on his concept-script was thus aimed at showing how it *works*, not what it *is*, leaving the task of showing its philosophical significance to philosophers. To make his exposition as plain as possible, he made extensive use of various metaphors and other figures of speech that, instead of being duly considered simply as rhetorical devices, were taken as his doing philosophy proper. This has led to the rise of the putatively Fregean "Compositionality Principle" quite despite Frege's own statements to the opposite effect, and also to the complete blindness to the factual referencelessness of Frege's system (which just could not have been otherwise, given Frege's stance was that of *Platonic* realism), disclosed in the first part of this study. Making the due distinction between the rhetorical verbiage and the deeper philosophical content of Frege's work renders it possible to present the real motif behind Frege's introduction of the sign of generality into his concept-script, namely his appreciation of the dual nature of *any* expression with regard to its logical content and, thus, of the fundamental importance, for Frege, of introducing special elements into his concept-script to make this duality explicit. Taking Frege's conception that served as a philosophical foundation of his use of the sign of the quantifier, along with the aptly altered use of his original logical formalism as a starting point, Frege's conception of logical duality gets due extension and further elaboration with the help of complementary ideas, not just from Kant, but also from Charles Sanders Peirce (Type-Token dichotomy and, broader, Peirce's system of philosophical categories), Thoralf Skolem (Skolem function), Ray Jackendoff, Zeno Vendler, Solomon Feferman, Kurt Gödel, Noam Chomsky and others.

Subjecting the core of Frege's thought to further consideration in the context of Peirce's nomenclature and divisions of dyadic relations, as well as Kant's distinction between Reason and Sensibility leads to the disclosure of the dual nature of causal relations in conventional signs (as opposed to natural signs) and the corresponding duality of Peirce's Interpretants, which brings closer to the understanding of the nature of Mind relative to Kant's basic distinction between Reason and Sensibility, the latter being parallel to the Mind *modulo* its human existence "here and now" while the former, to the same Mind *modulo* its being Replica (in Peirce's sense) of the collective mind of the species. Finally, the implementation of the full array of Peirce's philosophical categories is shown as a due method for revealing the inner structure of representations of quantity and thus the true essence of the categories of quantity as the logical requisites of cognition.

The attained results form the natural basis for the complementary part of the study, focused on the philosophical consequences of the application of the axiomatic method in linguistics. The analysis of quantificational facts in natural language with special reference to their derivational history as it is rendered in the theory of generative-transformational grammar provides ground to argue that the generally accepted view according to which the status and interpretation of Gödel's results have no particular bearing on linguistics is basically wrong and, thus, Chomsky's claim (that grammars are not machines), entirely unfounded. Moreover, having been shown to fall under the purview of Gödel's Incompleteness Theorem, the axiomatic formulation of linguistics appears to provide a suitable diagonalization method, analogous to that used by Gödel in the proof of his Theorem, that is indispensable for shedding the proper light on the deeper philosophical consequences of Gödel's Incompleteness Theorem. Ultimately, these philosophical consequences have been shown to be able to provide a crucial philosophical test case for Kant's doctrine of the synthetic *a priori*.

REFERENCES

Aldridge, M. V. 1982. *English Quantifiers. A Study of Quantifying Expressions in Linguistic Science and Modern English Usage*. Amersham: Avebury.
Allison, Henry E. 2006. Kant's Transcendental Idealism. In Graham Bird, ed., *A Companion to Kant*. Oxford: Blackwell.
Ayer, A. J. 1968. *The Origins of Pragmatism*. London: Macmillan.
Bach, Emmon. 2003. *Linguistic Universals and Particulars*. XVII International Congress of Linguists. Prague, 26 July 2003.
Baker, Gordon P., and Peter M. S. Hacker. 1984. *Language, Sense and Nonsense: A Critical Investigation into Modern Theories of Language*. Oxford: Basil Blackwell.
——. 1984. *Frege: Logical Excavations*. Oxford: Basil Blackwell.
Bańczerowski, Jerzy. 2006. The axiomatic method in 20th-century European linguistics. In Sylvain Auroux, E. F. K. Koerner, Hans-Josef Niederehe, Kees Versteegh, eds., *History of the Language Sciences*. Berlin and New York: Walter de Gruyter.
Barwise, Jon. 1976. Some Applications of Henkin Quantifiers. In *Israel Journal of Mathematics*, 25: 47–80.
Bell, David. 1979. *Frege's Theory of Judgement*. Oxford.
Bennett, Michael. 1979. Mass Nouns and Mass Terms in Montague Grammar. In *Linguistics, Philosophy and Montague Grammar*. Austin, TX: University of Texas Press.
Borgmann, Albert. 1978. Heidegger and Symbolic Logic. In *Heidegger and Modern Philosophy*, M. Murray, ed. New Haven and London: Yale University Press.
Brandom, Robert B. 1986. Frege's Technical Concepts. In Haaparanta, Leila, and Jaakko Hintikka, eds., *Frege Synthesized: Essays on the Philosophical and Foundational Work of Gottlob Frege*. Dordrecht: D. Reidel Publishing Company, 253–295.
Bunt, H. 1985. *Mass Terms and Model-theoretic Semantics*. Cambridge: Cambridge University Press.
Burge, Tyler. 1986. Frege on Truth. In Haaparanta, Leila, and Jaakko Hintikka, eds., *Frege Synthesized: Essays on the Philosophical and Foundational Work of Gottlob Frege*. Dordrecht: D. Reidel Publishing Company, 97–154.
Carden, Guy. 1976. *English Quantifiers: Logical Structure and Linguistic Variation*. New York, NY: Academic Press.
Chellas, B. F. 1979. Quantity and Quantification. In *Mass Terms, some Philosophical Problems*.
Chomsky, Noam. 1957. *Syntactic Structures*. The Hague: Mouton.
——. 1965. *Aspects of the Theory of Syntax*. Cambridge, Mass: MIT Press.
——. 1980. *Rules and Representations*. New York: Columbia University Press.
——. 1988. *Language and the Problems of Knowledge*. MIT Press.
——. 2001. Derivation by phase. In Michael Kenstowicz, ed., *Ken Hale: A Life in Language* (Current Studies in Linguistics 36), 1–52. Cambridge, MA: MIT Press.
——. 2006. *Language and Mind*. Cambridge, UK: Cambridge University Press.
Collins, John. 2009. A Question of Irresponsibility: Postal, Chomsky, and Gödel. *Biolinguistics* 3.1.
Currie, G. 1982. *Frege: An Introduction to his Philosophy*. Brighton: The Harvester Press.
Dean, Janet. 1966. *Determiners and Relative Clauses*. M.I.T., unpublished.
Debrock, G. 1996. La información y el regalo de Peirce al mundo. In *Anuario Filosófico* 29, 1331–1344.
Dummett, Michael. 1973. *Frege: Philosophy of Language*. London: Duckworth.
——. 1981. *The Interpretation of Frege's Philosophy*. Cambridge.

Enderton, Herbert. 1970. Finite Partially-Ordered Quantifiers. In *Zeitschrift für Mathematische Logik und Grundlagen der Mathematik*, 16: 393–397.

Faris, J. A. 1964. *Quantification Theory*. London: Routledge and Kegan Paul.

Feferman, Solomon. 2006. Are There Absolutely Unsolvable Problems? Gödel's Dichotomy. *Philosophia Mathematica* (III), Vol. 14 No. 2, 134–152.

——. 2007. Gödel, Nagel, minds and machines. Ernest Nagel Lecture, Columbia University, Sept. 27, 2007.

Fillmore, C., and D. T. Langendoen, eds. 1971. *Studies in Linguistic Semantics*. New York: Holt, Rinehart and Winston.

Føllesdal, Dagfinn. 2004. *Referential Opacity and Modal Logic*. London: Routledge.

Fraenkel A. A., Y. Bar-Hillel, A. Levy. 1973. *Foundations of Set Theory*. Amsterdam: North-Holland Publishing Company.

Frege, Gottlob. 1879. *Begriffsschrift, eine der arithmetischen nachgebildete Formelsprache des reinen Denkens*. Halle: L. Nebert.

——. 1882. Über die wissenschaftliche Berechtigung einer Begriffsschrift. Reprinted in I. Angelelli (ed.), *Begriffsschrift*, Hildesheim, 1964.

——. 1882–3. Über den Zweck der Begriffsschrift. *Jenaische Zeitschrift für Naturwissenschaft*, 16: 1–10.

——. 1884. *Die Grundlagen der Arithmetik: eine logisch-mathematische Untersuchungen uber den Begriff der Zahl*. Breslau.

——. 1891. *Funktion und Begriff*. Jena: H. Pohle.

——. 1892. Über Sinn und Bedeutung. *Zeitschrift für Philosophie und philosophische Kritik*, 100: 25–50.

——. 1892. Über Begriff und Gegenstand. *Vierteljahrschrift fur wissenschaftliche Philosophie*, 16: 192–205.

——. 1894. Review of E. Husserl, *Philosophie der Arithmetik*. *Zeitschrift für Philosophie und philosophische Kritik*, 103: 313–32.

——. 1895. Kritische Beleuchtung einiger Punkte in E. Schröders Vorlesungen über die Algebra der Logik. In *Archiv für systematische Philosophie*, 1: 433–56.

——. 1904. Was ist eine Funktion. In *Festschrift* Ludwig Boltzmann *gewidmet zum sechzigsten Geburtstage, 20. Februar 1902*. Leipzig: A. Barth, 656–66.

——. 1967. *Kleine Schriften*. I. Angelelli, ed. Hildesheim.

——. 1972. *Conceptual Notation and Related Articles*. Oxford: Clarendon Press.

——. 1979. *Posthumous Writings*. Hermes, H., F. Kambartel, F. Kaulbach, eds. Oxford: Basil Blackwell.

——. 1980. Philosophical and Mathematical Correspondence, Gabriel G., H. Hermes, F. Kambartel, C. Thiel, A. Veraart, eds., Chicago: The University of Chicago Press.

Garver, N. 1959. Wittgenstein on Private Language. *Philosophy and Phenomenological Research*, XX.

Geach, P., and M. Black, eds. 1952. *Translations from the Philosophical Writings of Gottlob Frege*. Oxford: Basil Blackwell.

Gödel, Kurt. 1931. Über formal unentscheidbare Sätze der *Principia Mathematica* und verwandter Systeme I', *Monatshefte für Mathematik und Physik* 38: 173–198. Reprinted and translated in Gödel (1986, 144–195).

——. 1986. *Collected Works Volume I*, S. Feferman, et al., eds., Oxford University Press.

——. 1951. Some basic theorems on the foundations of mathematics and their implications. In *Kurt Gödel, Collected Works, Volume III*. Oxford University Press, 1995, 304–323.

——. 1961. The modern development of the foundations of mathematics in the light of philosophy. In *Kurt Gödel, Collected Works, Volume III*. Oxford University Press, 1995.

——. 1995. *Collected Works, Vol. III: Unpublished Essays and Lectures*, S. Feferman, et al., eds., Oxford University Press.

Goldfarb, W. D. 1979. Logic in the Twenties: the Nature of the Quantifier. *Journal of Symbolic Logic*, 44: 351–368.

Gomperz, Heinrich. 1908. *Weltanschauungslehre*, vol. 2: *Noologie*, pt. 1: *Einleitung und Semasiologie*. Jena: Diederichs.

Guyer, Paul. 2010. The Deduction of the Categories: The Metaphysical and Transcendental Deductions. In Guyer, Paul, ed., *The Cambridge Companion to Kant's Critique of Pure Reason*. Cambridge: Cambridge University Press, 118–150.
Haaparanta, Leila, and Jaakko Hintikka, eds. 1986. *Frege Synthesized: Essays on the Philosophical and Foundational Work of Gottlob Frege*. Dordrecht: D. Reidel Publishing Company.
Hall, Barbara. 1962. *All About Predeterminers*. M.I.T., unpublished.
Hanna, Robert. 2001. *Kant and the Foundations of Analytic Philosophy*. Oxford: Clarendon/ Oxford Univ. Press.
——. 2008. Kant in the Twentieth Century. In Dermont Moran, ed., *The Routledge Companion to Twentieth-Century Philosophy*. London: Routledge.
Harris, Roy, and Talbot J. Taylor, 1997. *Landmarks in Linguistic Thought I: The Western Tradition from Socrates to Saussure*.—2nd ed. London: Routledge.
Heidegger, Martin. 1912. Neuere Forschungen über Logik. *Literarische Rundschau für das katholische Deutschland*, vol. 38.
——. 1961. *Nietzsche*. Pfullingen.
——. 1962. *Die Frage nach dem Ding*. Tübingen.
Henkin, Leon. Some Remarks on Infinitely Long Formulas. In *Infinitistic Methods*, Warsaw, 1959, 167–183.
Hintikka, Jaakko. 1974. Quantifiers vs. Quantification Theory. *Linguistic Inquiry* 5.
——. 1976. Quantifiers in Logic and Quantifiers in Natural Languages. In S. Körner, ed., *Philosophy of Logic*. Oxford: Blackwell.
——. 1996. *The Principles of Mathematics Revisited*. Cambridge: Cambridge University Press.
Hintikka, Jaakko, and Gabriel Sandu. 1994. Uses and misuses of Frege's ideas. *The Monist* 77: 278–293.
Hintikka, Jaakko, and Gabriel Sandu. 1997. Game-Theoretical Semantics. In J. van Benthem, and A. ter Meulen, eds., *Handbook of Logic and Language*, 361–410. Amsterdam: Elsevier.
Hogg, Richard M. 1972. Quantifiers and Possessives. *Lingua* 30.
Hornstein, N. 1984. *Logic as Grammar. An Approach to Meaning in Natural Language*. Cambridge, Mass.: MIT Press.
Hutton, Christopher M. 1990. *Abstraction and Instance: The Type-Token Relation in Linguistic Theory*. Oxford: Pergamon Press.
Jackendoff, Ray S. 1968. Quantifiers in English. *Foundations of Language* 4: 422–442.
Jespersen, O. 1924. *The Philosophy of Grammar*. London: George Allen and Unwin.
——. 1969. *Analytic Syntax*. New York: Holt, Rinehart and Winston.
Jevons, W. Stanley. 1897. *Elementary Lessons in Logic: Deductive and Inductive*. London.
Kant, Immanuel. 1965. *Critique of Pure Reason*, trans. N. Kemp Smith. New York: St Martin's.
——. 1967. *Philosophical Correspondence, 1759–99*. Chicago: The University of Chicago Press.
Kelman, James. 1988. *A reading from Noam Chomsky and the Scottish Tradition in the Philosophy of Common Sense*. Unpublished.
Kneale, William and Martha Kneale. 1962. *The Development of Logic*. Oxford: Clarendon Press.
Körner, S. ed. 1976. *Philosophy of Logic*. Oxford: Blackwell.
Kreisel, Georg. 1972. Which number-theoretic problems can be solved in recursive progressions on Π^1_1 paths through O?, *J. Symbolic Logic* 37: 311–334.
Krynicki, M., M. Mostowski. 1990. Henkin Quantifiers. In Krynicki, M., M. Mostowski, and L. W. Szczerba, eds., *Quantifiers: Logics, Models and Computation*. Dordrecht: Kluwer Academic Publishers, 193–259.
Krynicki, M., M. Mostowski, and L. W. Szczerba, eds. 1990. *Quantifiers: Logics, Models and Computation*. Dordrecht: Kluwer Academic Publishers.
Ladusaw, William A. 1983. Logical form and conditions on grammaticality. *Linguistics and Philosophy*, 6:3, 373–392.

Lakoff, G. 1971. On Generative Semantics. In Steinberg and Jakobovits, eds., *Semantics: An Interdisciplinary Reader*. Cambridge: Cambridge University Press.
Laycock, H. 1975. Theories of Matter, *Synthese* 31: 411–442.
——. 1998. Words without objects. *Principia*, vol. 2 no. 2, 147–182.
Lee, D. A. 1971. Quantifiers and Identity in Relativisation. *Lingua* 27.
——. 1972. Reply to „Quantifiers and Possessives." *Lingua* 30.
LePore, E., ed. 1987. *New Directions in Semantics*. London: Academic Press.
Levelt, Willem J. M. 2008. *An Introduction to the Theory of Formal Languages and Automata*. John Benjamins Publishing Company.
Levine, Alexander T. 1999. Scientific Progress and the Fregean Legacy, *Mind & Language*, 14:3, 263–290.
Lieb, Hans-Heinrich. 1981. Das ‚semiotische Dreieck' bei Ogden und Richards: eine Neuformuliering des Zeichenmodells von Aristoteles. In Geckeler, Horst, et al., eds., *Logos Semanticos*, Berlin: de Gruyter, 137–56.
Lucas, J. R. 1964. Minds, Machines and Gödel. In Alan R. Anderson, ed., *Minds and Machines*. Englewood Cliffs, NJ: Prentice Hall.
——. 2002. The Implications of Gödels Theorem. In J. R. Lucas, *Conceptual Roots of Mathematics*. London: Routledge, 199–223.
Lyons, John. 1977. *Semantics*. 2 vols. Cambridge: Cambridge University Press.
Malmkjær, Kirsten, ed. 1991. *The Linguistics Encyclopedia*. London and New York: Routledge.
Manaster-Ramer, Alexis, ed. 1987. *Mathematics of Language*, Amsterdam: John Benjamins.
Martin-Löf, Per. 1995. Verificationism then and now. In W. De Pauli-Schimanovich, et al., eds., *The Foundational Debate*. Dordrecht: Kluwer, 187–196.
May, R. 1985. *Logical Form. Its Structure and Derivation*. Cambridge, Mass.: MIT Press.
——. 1987. Logical Form as a Level of Linguistic Representation. In E. LePore and B. Richards, eds., *New Directions in Semantics*. London: Academic Press, 305–336.
Marr, D. 1982. *Vision: A computational investigation into the human representation and processing of visual information*, W.H. Freeman publ.
McCawley, James D. 1993. Everything That Linguists Have Always Wanted to Know about Logic—But Were Ashamed to Ask,—2nd ed. Chicago: The University of Chicago Press.
Mill, John Stuart. (1843) 1925. *A System of Logic*. London: Longman, Green, and Co.
Mitchell, D. 1962. *An Introduction to Logic*. London: Hutchinson and Co.
Mołczanow, A. 1994. Set Reference Relationships and the Phrasal Syntax of Quantifiers in English, *Studia Anglica Posnaniensia* XXIX: 73–90.
——. 2002. Quantification and Inference, *The Monist*, 85:4, 535–554.
——. 2003. Logical Entailment on Multidimensional Branching Quantifier Representations, *Annales UMCS Philosophia-Sociologia*, vol. XXVIII, Lublin, 129–136.
Moltmann, Friederike. 2010. *The Number of Planets: A Number-Referring term?*, volume on abstractionism, edited by M. Rossberg and P. Ebert, under review.
Nöth, Winfried. 1990. *Handbook of Semiotics*. Bloomington: Indiana University Press.
Nubiola, Jaime. 2002. Peirce on Complexity. In W. Schmitz, ed., *Proceedings of the 7th International Congress of the IASS-AIS*, Dresden: Thelem, 11–23.
Ogden, Charles Kay, and Ivor Armstrong Richards (1923) 1946. *The Meaning of Meaning*. New York: Harcourt.
Orenstein, Alex. 1983. Towards a Philosophical Classification of Quantifiers. In A. Orenstein, and R. Stern, eds., *Developments in Semantics*. New York: Haven Publications.
Parker, K. A. 1998. *The Continuity of Peirce's Thought*. Nashville, TN: Vanderbilt University Press.
Partee, B. H. 1971. On the Requirement that Transformations Preserve Meaning. In C. Fillmore, and D. T. Langendoen, eds., *Studies in Linguistic Semantics*. New York: Holt, Rinehart and Winston.
Peirce, Charles Sanders. 1960. *Collected Papers of Charles Sanders Peirce*, Ch. Hartshorne, P. Weiss, A. W. Burks, eds., Cambridge, MA: Harvard University Press.

Pelczar, Michael. 2000. Wittgensteinian Semantics. *Noûs*, 34:4, 483–516.
Pelletier, F. J., ed. 1979. *Mass Terms, some Philosophical Problems*. Dordrecht: Reidel.
Pharies, David A. 1985. *Charles S. Peirce and the Linguistic Sign*. Amsterdam: John Benjamins.
Postal, Paul M. 2009. The Incoherence of Chomsky's 'Biolinguistic' Ontology. *Biolinguistics* 3.1.
Ross, Kelley L. 1985. *The Origin of Value in a Transcendent Function*. PhD diss., University of Texas at Austin.
Quine, W. V. O. 1940. *Mathematical Logic*. New York: W. W. Norton & Company.
——. 1970. *Philosophy of Logic*. Englewood Cliffs: Prentice-Hall.
——. 1972. *Methods of Logic*. London: Routledge & Kegan Paul.
Rorty, Richard. 1982. *Consequences of Pragmatism*. Minneapolis: University of Minnesota Press.
Russell, B. 1903. *The Principles of Mathematics*. Cambridge: Cambridge University Press.
——. 1946. *History of Western Philosophy*. London: George Allen and Unwin.
Saussure, Ferdinand de. 1959. *Course in General Linguistics*. New York: The Philosophical Library, Inc.
Seuren, P. A. M. 1969. *Operators and Nucleus. A Contribution to the Theory of Grammar*. Cambridge: Cambridge University Press.
Silverman, K. 1983. *The Subject of Semiotics*. New York: Oxford University Press.
Sluga, Hans D. 1980. *Gottlob Frege*. London: Routledge & Kegan Paul.
Steinberg, D. D., and L. A. Jakobovits, eds. 1971. *Semantics: An Interdisciplinary Reader*. Cambridge: Cambridge University Press.
Thomason, Richmond H. 1987. Remarks on Linguistic Semantics. In A. Manaster-Ramer, ed., *Mathematics of Language*. Amsterdam: John Benjamins.
Thrane, Torben. 1980. *Referential-Semantic Analysis: Aspects of a Theory of Linguistic Reference*. Cambridge: Cambridge University Press.
Tiles, Mary. 2004. Kant: From General to Transcendental Logic. In Dov. M. Gabbay and John Woods, eds., *Handbook of the History of Logic: Volume 3, The Rise of Modern Logic: From Leibniz to Frege*. Amsterdam: Elsevier.
Trendelenburg, A. 1867. On Leibniz's Project of a Universal Characteristic. *Historische Beiträge zur Philosophie*, vol. 3, Berlin.
Vendler, Z. 1967. Each and Every, Any and All. In Z.Vendler, *Linguistics in Philosophy*. Ithaca, N.Y.: Cornell University Press.
Walkoe, W. 1970. Finite Partially Ordered Quantification. In *Journal of Symbolic Logic* 35: 535–50.
Westerståhl, Dag. 1986. Quantifiers in Formal and Natural Languages. *CSLI Publications, Report No. CSLI-86-55*, Stanford.
Wiggins, D. 1980. 'Most' and 'All': Some Comments on a Familiar Programme, and on the Logical Form of Quantified Sentences. In M.Platts, ed., *Reference, Truth and Reality: Essays on the Philosophy of Language*. London: Routledge and Kegan Paul.
Wittgenstein, L. 1922. *Tractatus Logico-Philosophicus*, London – New York.
——. 1980. *Remarks on the Philosophy of Psychology*. Chicago: Chicago University Press.
Wright, Georg Henrik von. 1951. *An Essay in Modal Logic*. Amsterdam: North-Holland Pub. Co.
Zimmerman, D. W. 1995. Theories of Masses and Problems of Constitution. *The Philosophical Review* 104: 53–110.

INDEX

www.ingramcontent.com/pod-product-compliance
Lightning Source LLC
La Vergne TN
LVHW081351011026
849161LV00013BA/2785